AF785114

ELSEVIER

VOLUME 8, ISSUE 7
DECEMBER 1998
ISBN 0-444-10032-6

AIR POLLUTION CONSULTANT

Quick Reference Guide

The Air Pollution Consultant is published seven times per annum in the months of January, March, May, July, September, November, and December by Elsevier Science Inc., 655 Avenue of the Americas, New York, New York 10010-5107. The annual subscription rate is $745 within the USA and $825 outside the USA. Subscriptions to individual issues are available separately. To place an order or for customer service, call (303) 470-1900 ext. 12.

Postmaster: Please send address changes to *The Air Pollution Consultant*, Elsevier Science Inc., 655 Avenue of the Americas, New York, NY 10010. Claims for missing issues can be honored only up to three months for domestic addresses and six months for foreign addresses. Duplicate copies will not be sent to replace ones undelivered through failure to notify Elsevier Science Inc. of change of address.

All inquiries regarding the technical content of this journal should be directed to: Vicki A. Dean, Editor, *The Air Pollution Consultant*, Elsevier Science Inc., 8773 South Ridgeline Blvd, Highlands Ranch, CO 80126-2329, (303) 470-1900, Fax (303) 470-5119, v.a.dean@elsevier.com.

Table of Contents

©Elsevier Science Inc. 0-444-10032-6/98/$19.00+$5.50

1 Introduction

How often have you needed to quickly refer to a key list of Clean Air Act (CAA) chemicals or deadlines only to find that you had spent an hour or more searching for it in the *Code of Federal Regulations* (*CFR*), your files, or back issues of *The Air Pollution Consultant* (*APC*)? Other times, you may have needed to contact EPA or your local regulatory agency to obtain information or resolve a specific compliance question. Unfortunately, it may have taken a number of phone calls and considerable effort just to contact the right office, let alone a person who could help you. Finally, most of us are aware that more and more CAA-related information is available via the Internet, but how do we access that information and, more importantly, reduce the signal-to-noise ratio?

The Purpose of This Guide

This revised *APC Quick Reference Guide* is designed to help you easily and accurately answer important, everyday questions related to the CAA and stationary sources. The chapters that follow provide rapid access to key CAA information that is often difficult to find. Such information includes important regulatory lists (e.g., hazardous air pollutants, ozone-depleting substances), *CFR* references for specific programs, logic diagrams to help with CAA program applicability determinations, lists of sources regulated under specific programs, regulatory compliance deadlines, and much more.

The guide also includes an extensive list of CAA-related resources, road maps to key EPA Web sites, and a directory of important Internet sites. Information on how to access regulations, EPA publications, technical guidance on a number of topics, environmental law libraries, etc. is provided, as well as comprehensive lists of EPA and state and local regulatory agency contacts.

Contents

The following topics are covered in this guide book:

- Chapter 2—General CAA Regulatory Information
- Chapter 3—The National Ambient Air Quality Standards and New Source Review Programs
- Chapter 4—New Source Performance Standards
- Chapter 5—Hazardous Air Pollutant Control Programs
- Chapter 6—The Operating Permit Program
- Chapter 7—The Stratospheric Ozone Protection Program
- Chapter 8—The Acid Rain Program
- Chapter 9—Where to Go for More Information
- Chapter 10—Road Maps for Navigating Key EPA Web Sites
- Chapter 11—Directory of CAA-Related Web Sites
- Chapter 12—Professional Toolbox
- Chapter 13—Keyword Index
- Chapter 14—Glossary of Terms

New Edition Provides Updated Information and Additional Features

This second edition of the *APC Quick Reference Guide* has been updated to reflect all regulatory changes through September 1998. In addition to providing current information, we have included several new features. To help navigate around the ins and outs of EPA's growing Internet presence, Chapter 10 provides you with road maps of several key agency Web sites. The Internet directory in Chapter 11 has also been expanded and updated, Web site addresses have been changed where necessary, and nearly *100 new sites* have been added. Finally, we have rounded out this guide with a compilation of useful reference conversion factors and facts in the Toolbox (Chapter 12).

How to Use the *APC Quick Reference Guide*

In an effort to make this guide easy to use, we have organized the bulk of the information we present by regulatory program (e.g., the operating permit program). Chapter 2 presents an introductory overview of CAA programs, regulations, and related rulemakings. Chapters 3 through 8 contain key regulatory lists and diagrams for specific regulatory programs. Chapters 9 through 11 summarize resources that are available for obtaining more information. As described above, Chapter 12 presents an array of useful but hard-to-find equations, conversion factors, physical constants, and other technical tidbits. Finally, we round out the *APC Quick Reference Guide* with a keyword index and extensive glossary of terms.

Spread amongst the various chapters of this guide are a number of features designed to present information in especially useful formats. These include the following:

- *Logic diagrams*—A number of logic diagrams are provided to guide you step-by-step through key CAA regulations. The diagrams can help you determine if you are subject to the regulations and, if so, what you must do to comply.

- *Regulatory compliance calendars*—Where applicable, our chapters dealing with specific regulatory programs include a regulatory compliance calendar that shows *when* a regulated source must comply with a given requirement. This feature is especially important for the national emission standards for hazardous air pollutants (NESHAP) program, because the compliance dates for several maximum achievable control technology (MACT) standards fall during 1999 (e.g., printing and publishing, and offsite waste and recovery MACT standards).

- *APC cross-references*—In our seven years of publishing this journal, we have reviewed essentially all of the regulations implementing stationary source controls under the Clean Air Act Amendments of 1990. So that our readers can easily find these articles in past issues of *APC* and thus obtain more information about a specific CAA program, we have included *APC* cross-reference information in Chapters 3 through 8. For additional, more detailed past article information, you can also consult the annual cumulative keyword index.

- *Road maps to key EPA Web sites*—Getting the information you need via the Internet will be much easier using our new road maps to key EPA Web sites (see Chapter 10). The road maps provide in-depth details of some of EPA's most important CAA-related Web sites and where they can be accessed—saving you research time and effort.

- *Internet directory*—Perhaps the most valuable section of this guide, our Internet directory, lists Web sites that are increasingly important sources of CAA information. The directory is organized and presented in two ways: 1) Web site names (e.g., Air & Waste Management Association, EPA's Environmental Appeals Board), and 2) key word-referenced Web sites (e.g., monitoring, emission factors). Using this directory, you can quickly access a plethora of information directly from a computer at work or home.

- *Professional toolbox*—Helpful facts, rules-of-thumb, and conversion factors (just like all those dog-eared copies in your briefcase) have been compiled in Chapter 12.

- *Keyword index*—A keyword index is included in Chapter 13 to provide easy access to the contents of this guide by specific topics of interest. APC

2 General CAA Regulatory Information

Wading through the *CFR* to find items pertaining to a specific CAA program can be a tedious task, especially considering that regulations are scattered throughout numerous volumes. To make this task a bit easier, we have compiled a master table of contents to the CAA-related *CFR*. The list, which is contained in Table 2-1 (pages 2.2–2.12), provides just enough detail to guide you in the right direction, but not so much that it is overwhelming.

Given the complexities of CAA compliance, simply having familiarity with existing regulations is not enough. EPA is constantly promulgating new regulations that impact the regulated community. Consequently, it is a good idea to know which regulations EPA is planning to issue in the near future. That way, you can anticipate regulations that may affect your facility, business, or client and allow yourself more time to 1) comment on important proposed rule issues, and 2) prepare for compliance with final rules.

EPA's Schedule of Planned CAA Rulemakings

Twice each year, EPA issues a regulatory agenda that summarizes upcoming CAA-related rulemakings and provides information on when the agency expects to propose and finalize each rule. Using the last version of the regulatory agenda, issued on April 27, 1998 (63 *FR* 22602), we have taken EPA's offering a step further by linking agency rulemakings to regulatory analyses presented in *APC*. The result, which is presented in Table 2-2 (pages 2.13–2.27), is described below. (Note that EPA's semiannual regulatory agenda is published in the *Federal Register* [*FR*] and is available on the Internet at http://www.epa.gov/fedrgstr/unified.htm.)

The rules included in Table 2-2 are grouped by stage of development, specifically 1) rules under consideration that have not been formally announced, 2) those that have been announced but not yet issued as proposed rules, 3) those that have been proposed but not yet finalized, and 4) those that have been finalized in the months since EPA published the last regulatory agenda (i.e., between late April and October 1998). Within each of the four categories, entries are first sorted by *CFR* part number affected, and then by the date of the next scheduled regulatory action.

For each entry, the first two columns provide a brief summary of the anticipated change and the parts of the regulations that will be affected. The next two columns list proposed and/or final actions and the dates on which they occurred or are scheduled to occur. The actual completion date for a given action is shown in boldface type with the *FR* reference given below the action date.

The legal deadline for each activity is listed in a separate column under either the proposed or final action column. An "S" or "J" behind the date signifies whether it is a statutory or judicial deadline (i.e., court order). (Warning: EPA often misses self-established and statutory deadlines; however, the agency usually meets judicial deadlines.) As the new century approaches, entries for the year 2000 and beyond are beginning to appear. We list these dates in the same manner as EPA; for example, "11/00" signifies November 2000.

Finally, if an action has been reviewed in *APC*, the volume, issue, and page number of the review are listed in the columns labeled "*APC* Review." For example, "8:5, 2.2" means that a review of this rulemaking begins on page 2.2 of Volume 8, Issue 5 of *APC*.

TABLE 2-1
Master Table of Contents for CAA-Related *Code of Federal Regulations* Entries

Part/Subpart	*Title*
Part 49	Tribal Clean Air Act Authority
Part 50	National Primary and Secondary Ambient Air Quality Standards
Part 51	Requirements for Preparation, Adoption, and Submittal of Implementation Plans
A–E	• [Reserved]
F	• Procedural Requirements
G	• Control Strategy
H	• Prevention of Air Pollution Emergency Episodes
I	• Review of New Sources and Modifications
J	• Ambient Air Quality Surveillance
K	• Source Surveillance
L	• Legal Authority
M	• Intergovernmental Consultation
N	• Compliance Schedules
O	• Miscellaneous Plan Content Requirements
P	• Protection of Visibility
Q	• Reports
R	• Extensions
S	• Inspection/Maintenance Program Requirements
T	• Conformity to State or Federal Implementation Plans of Transportation Plans, Programs, and Projects Developed, Funded or Approved Under Title 23 U.S.C. or the Federal Transit Act
U	• Economic Incentive Programs
W	• Determining Conformity of General Federal Actions to State or Federal Implementation Plans
Part 52	Approval and Promulgation of Implementation Plans
A	• General Provisions
B–ZZ	• Alabama . . . Wyoming (i.e., one subpart with SIP information for each state, including the District of Columbia)
AAA–DDD	• Guam, Puerto Rico, Virgin Islands, and American Samoa (SIP information for each respective area)
EEE	• Approval and Promulgation of Plans
FFF	• Commonwealth of the Northern Mariana Islands (SIP information)
Part 53	Ambient Air Monitoring Reference and Equivalent Methods
A	• General Provisions
B	• Procedures for Testing Performance Characteristics of Automated Methods for SO_2, CO, O_3, and NO_2
C	• Procedures for Determining Comparability Between Candidate Methods and Reference Methods
D	• Procedures for Testing Performance Characteristics of Methods for PM-10

TABLE 2-1

Master Table of Contents for CAA-Related *Code of Federal Regulations* Entries—Continued

Part/Subpart	*Title*
Part 54	Prior Notice of Citizen Suits
Part 55	Outer Continental Shelf Air Regulations
Part 56	Regional Consistency
Part 57	Primary Nonferrous Smelter Orders
A	• General
B	• The Application and the NSO Process
C	• Constant Controls and Related Requirements
D	• Supplementary Control System Requirements
E	• Fugitive Emission Evaluation and Control
F	• Research and Development Requirements
G	• Compliance Schedule Requirements
H	• Waiver of Interim Requirement for Use of Continuous Emission Reduction Technology
Part 58	Ambient Air Quality Surveillance
A	• General Provisions
B	• Monitoring Criteria
C	• State and Local Air Monitoring Stations
D	• National Air Monitoring Stations
E	• Photochemical Assessment Monitoring Stations
F	• Air Quality Index Reporting
G	• Federal Monitoring
Part 59	Consumer and Commercial Products
A	• [Reserved]
B	• Automotive Refinish Coatings
C	• Consumer Products
D	• Architectural Coatings
Part 60	Standards of Performance for New Stationary Sources
A	• General Provisions
B	• Adoption and Submittal of State Plans for Designated Facilities
C	• Emission Guidelines and Compliance Times
Ca	• [Reserved]
Cb	• Emission Guidelines and Compliance Times for Municipal Waste Combustors That Are Constructed on or Before December 19, 1995
Cc	• Emission Guidelines and Compliance Times for Municipal Solid Waste Landfills
Cd	• Emission Guidelines and Compliance Times for Sulfuric Acid Production Units
Ce	• Emission Guidelines and Compliance Times for Hospital/Medical/Infectious Waste Incinerators

TABLE 2-1
Master Table of Contents for CAA-Related *Code of Federal Regulations* Entries—Continued

Part/Subpart	*Title*
Part 60—Continued	
D	• Standards of Performance for Fossil Fuel Fired-Steam Generators for Which Construction Is Commenced After August 17, 1971
Da	• Standards of Performance for Electric Utility Steam Generating Units for Which Construction Is Commenced After September 18, 1978
Db	• Standards of Performance for Industrial-Commercial-Institutional Steam Generating Units
Dc	• Standards of Performance for Small Industrial-Commercial-Institutional Steam Generating Units
E	• Standards of Performance for Incinerators
Ea	• Standards of Performance for Municipal Waste Combustors for Which Construction Is Commenced After December 20, 1989 and on or Before September 20, 1994
Eb	• Standards of Performance for Municipal Waste Combustors for Which Construction Is Commenced After September 20, 1994
Ec	• Standards of Performance for Hospital/Medical/Infectious Waste Incinerators for Which Construction Is Commenced After June 20, 1996
F	• Standards of Performance for Portland Cement Plants
G	• Standards of Performance for Nitric Acid Plants
H	• Standards of Performance for Sulfuric Acid Plants
I	• Standards of Performance for Hot Mix Asphalt Facilities
J	• Standards of Performance for Petroleum Refineries
K	• Standards of Performance for Storage Vessels for Petroleum Liquids for Which Construction, Reconstruction, or Modification Commenced After June 11, 1973, and Prior to May 19, 1978
Ka	• Standards of Performance for Storage Vessels for Petroleum Liquids for Which Construction, Reconstruction, or Modification Commenced After May 18, 1978, and Prior to July 23, 1984
Kb	• Standards of Performance for Volatile Organic Liquid Storage Vessels (Including Petroleum Liquid Storage Vessels) for Which Construction, Reconstruction, or Modification Commenced after July 23, 1984
L	• Standards of Performance for Secondary Lead Smelters
M	• Standards of Performance for Secondary Brass and Bronze Production Plants
N	• Standards of Performance for Primary Emissions From Basic Oxygen Process Furnaces for Which Construction Is Commenced After June 11, 1973
Na	• Standards of Performance for Secondary Emissions From Basic Oxygen Process Steelmaking Facilities for Which Construction Is Commenced After January 20, 1983
O	• Standards of Performance for Sewage Treatment Plants
P	• Standards of Performance for Primary Copper Smelters
Q	• Standards of Performance for Primary Zinc Smelters
R	• Standards of Performance for Primary Lead Smelters
S	• Standards of Performance for Primary Aluminum Reduction Plants

TABLE 2-1

Master Table of Contents for CAA-Related *Code of Federal Regulations* Entries—Continued

Part/Subpart	*Title*
Part 60—Continued	
T	• Standards of Performance for the Phosphate Fertilizer Industry: Wet-Process Phosphoric Acid Plants
U	• Standards of Performance for the Phosphate Fertilizer Industry: Superphosphoric Acid Plants
V	• Standards of Performance for the Phosphate Fertilizer Industry: Diammonium Phosphate Plants
W	• Standards of Performance for the Phosphate Fertilizer Industry: Triple Superphosphate Plants
X	• Standards of Performance for the Phosphate Fertilizer Industry: Granular Triple Superphosphate Storage Facilities
Y	• Standards of Performance for Coal Preparation Plants
Z	• Standards of Performance for Ferroalloy Production Facilities
AA	• Standards of Performance for Steel Plants: Electric Arc Furnaces Constructed After October 21, 1974 and on or Before August 17, 1983
AAa	• Standards of Performance for Steel Plants: Electric Arc Furnaces and Argon-Oxygen Decarburization Vessels Constructed After August 7, 1983
BB	• Standards of Performance for Kraft Pulp Mills
CC	• Standards of Performance for Glass Manufacturing Plants
DD	• Standards of Performance for Grain Elevators
EE	• Standards of Performance for Surface Coating of Metal Furniture
FF	• [Reserved]
GG	• Standards of Performance for Stationary Gas Turbines
HH	• Standards of Performance for Lime Manufacturing Plants
KK	• Standards of Performance for Lead-Acid Battery Manufacturing Plants
LL	• Standards of Performance for Metallic Mineral Processing Plants
MM	• Standards of Performance for Automobile and Light-Duty Truck Surface Coating Operations
NN	• Standards of Performance for Phosphate Rock Plants
PP	• Standards of Performance for Ammonium Sulfate Manufacture
QQ	• Standards of Performance for the Graphic Arts Industry: Publication Rotogravure Printing
RR	• Standards of Performance for Pressure Sensitive Tape and Label Surface Coating Operations
SS	• Standards of Performance for Industrial Surface Coating: Large Appliances
TT	• Standards of Performance for Metal Coil Surface Coating
UU	• Standards of Performance for Asphalt Processing and Asphalt Roofing Manufacture
VV	• Standards of Performance for Equipment Leaks of VOC in the Synthetic Organic Chemicals Manufacturing Industry
WW	• Standards of Performance for the Beverage Can Surface Coating Industry
XX	• Standards of Performance for Bulk Gasoline Terminals
AAA	• Standards of Performance for New Residential Wood Heaters

TABLE 2-1
Master Table of Contents for CAA-Related *Code of Federal Regulations* Entries—Continued

Part/Subpart	*Title*
Part 60—Continued	
BBB	• Standards of Performance for the Rubber Tire Manufacturing Industry
CCC	• [Reserved]
DDD	• Standards of Performance for VOC Emissions From the Polymer Manufacturing Industry
EEE	• [Reserved]
FFF	• Standards of Performance for Flexible Vinyl and Urethane Coating and Printing
GGG	• Standards of Performance for Equipment Leaks of VOC in Petroleum Refineries
HHH	• Standards of Performance for Synthetic Fiber Production Facilities
III	• Standards of Performance for VOC Emissions From the Synthetic Organic Chemical Manufacturing Industry Air Oxidation Unit Processes
JJJ	• Standards of Performance for Petroleum Dry Cleaners
KKK	• Standards of Performance for Equipment Leaks of VOC From Onshore Natural Gas Processing Plants
LLL	• Standards of Performance for Onshore Natural Gas Processing; SO_2 Emissions
MMM	• [Reserved]
NNN	• Standards of Performance for Volatile Organic Compound Emissions From Synthetic Organic Chemical Manufacturing Industry Distillation Operations
OOO	• Standards of Performance for Nonmetallic Mineral Processing Plants
PPP	• Standard of Performance for Wool Fiberglass Insulation Manufacturing Plants
QQQ	• Standards of Performance for VOC Emissions From Petroleum Refinery Wastewater Systems
RRR	• Standards of Performance for Volatile Organic Compound Emissions From Synthetic Organic Chemical Manufacturing Industry Reactor Processes
SSS	• Standards of Performance for Magnetic Tape Coating Facilities
TTT	• Standards of Performance for Industrial Surface Coating: Surface Coating of Plastic Parts for Business Machines
UUU	• Standards of Performance for Calciners and Dryers in Mineral Industries
VVV	• Standards of Performance for Polymeric Coating of Supporting Substrates Facilities
WWW	• Standards of Performance for Municipal Solid Waste Landfills
Part 61	National Emission Standards for Hazardous Air Pollutants
A	• General Provisions
B	• National Emission Standards for Radon Emissions From Underground Uranium Mines
C	• National Emission Standard for Beryllium
D	• National Emission Standard for Beryllium Rocket Motor Firing
E	• National Emission Standard for Mercury
F	• National Emission Standard for Vinyl Chloride
G	• [Reserved]
H	• National Emission Standards for Emissions of Radionuclides Other Than Radon From Department of Energy Facilities

TABLE 2-1

Master Table of Contents for CAA-Related *Code of Federal Regulations* Entries—Continued

Part/Subpart	*Title*
Part 61—Continued	
I	• National Emission Standards for Radionuclide Emissions From Facilities Licensed by the Nuclear Regulatory Commission and Federal Facilities Not Covered by Subpart H
J	• National Emission Standard for Equipment Leaks (Fugitive Emission Sources) of Benzene
K	• National Emission Standards for Radionuclide Emissions From Elemental Phosphorus Plants
L	• National Emission Standard for Benzene Emissions From Coke By-Product Recovery Plants
M	• National Emission Standard for Asbestos
N	• National Emission Standard for Inorganic Arsenic Emissions From Glass Manufacturing Plants
O	• National Emission Standard for Inorganic Arsenic Emissions From Primary Copper Smelters
P	• National Emission Standard for Inorganic Arsenic Emissions From Arsenic Trioxide and Metallic Arsenic Production Facilities
Q	• National Emission Standards for Radon Emissions From Department of Energy Facilities
R	• National Emission Standards for Radon Emissions From Phosphogypsum Stacks
S	• [Reserved]
T	• National Emission Standards for Radon Emissions From the Disposal of Uranium Mill Tailings
U	• [Reserved]
V	• National Emission Standard for Equipment Leaks (Fugitive Emission Sources)
W	• National Emission Standards for Radon Emissions From Operating Mill Tailings
X	• [Reserved]
Y	• National Emission Standard for Benzene Emissions From Benzene Storage Vessels
Z–AA	• [Reserved]
BB	• National Emission Standard for Benzene Emissions From Benzene Transfer Operations
CC–EE	• [Reserved]
FF	• National Emission Standard for Benzene Waste Operations
Part 62	Approval and Promulgation of State Plans for Designated Facilities and Pollutants
A	• General Provisions
B–ZZ[1]	• Alabama . . . Wyoming [i.e., one subpart with CAA Section 111(d) plan information for each state, including District of Columbia][2]
AAA	• [Reserved]
BBB–CCC	• Puerto Rico and Virgin Islands [CAA Section 111(d) plans for each area]

TABLE 2-1
Master Table of Contents for CAA-Related *Code of Federal Regulations* Entries—Continued

Part/Subpart	*Title*
Part 63	National Emission Standards for Hazardous Air Pollutants for Source Categories
A	• General Provisions
B	• Requirements for Control Technology Determinations for Major Sources in Accordance With Clean Air Act Sections 112(g) and 112(j)
C	• List of Hazardous Air Pollutants, Petition Process, Lesser Quantity Designations, Source Category List
D	• Regulations Governing Compliance Extensions for Early Reductions of Hazardous Air Pollutants
E	• Approval of State Programs and Delegation of Federal Authorities
F	• National Emission Standards for Organic Hazardous Air Pollutants From the Synthetic Organic Chemical Manufacturing Industry
G	• National Emission Standards for Organic Hazardous Air Pollutants From Synthetic Organic Chemical Manufacturing Industry for Process Vents, Storage Vessels, Transfer Operations, and Wastewater
H	• National Emission Standards for Organic Hazardous Air Pollutants for Equipment Leaks
I	• National Emission Standards for Organic Hazardous Air Pollutants for Certain Processes Subject to the Negotiated Regulation for Equipment Leaks
J–K	• [Reserved]
L	• National Emission Standards for Coke Oven Batteries
M	• National Perchloroethylene Air Emission Standards for Dry Cleaning Facilities
N	• National Emission Standards for Chromium Emissions From Hard and Decorative Chromium Electroplating and Chromium Anodizing Tanks
O	• Ethylene Oxide Emission Standards for Sterilization Facilities
P	• [Reserved]
Q	• National Emission Standards for Hazardous Air Pollutants for Industrial Process Cooling Towers
R	• National Emission Standards for Gasoline Distribution Facilities (Bulk Gasoline Terminals and Pipeline Breakout Stations)
S	• National Emission Standards for Hazardous Air Pollutants from the Pulp and Paper Industry
T	• National Emission Standards for Halogenated Solvent Cleaning
U	• National Emission Standards for Hazardous Air Pollutant Emissions: Group I Polymers and Resins
V	• [Reserved]
W	• National Emission Standards for Hazardous Air Pollutants for Epoxy Resins Production and Non-Nylon Polyamides Production
X	• National Emission Standards for Hazardous Air Pollutants From Secondary Lead Smelting
Y	• National Emission Standards for Marine Tank Vessel Loading Operations
Z–BB	• [Reserved]
CC	• National Emission Standards for Hazardous Air Pollutants From Petroleum Refineries
DD	• National Emission Standards for Hazardous Air Pollutants From Offsite Waste and Recovery Operations

TABLE 2-1
Master Table of Contents for CAA-Related *Code of Federal Regulations* Entries—Continued

Part/Subpart	*Title*
Part 63—Continued	
EE	• National Emission Standards for Magnetic Tape Manufacturing Operations
FF	• [Reserved]
GG	• National Emission Standards for Aerospace Manufacturing and Rework Facilities
HH	• [Reserved]
II	• National Emission Standards for Shipbuilding and Ship Repair (Surface Coating)
JJ	• National Emission Standards for Wood Furniture Manufacturing Operations
KK	• National Emission Standards for the Printing and Publishing Industry
LL	• National Emission Standards for Hazardous Air Pollutants for Primary Aluminum Reduction Plants
MM–NN	• [Reserved]
OO	• National Emission Standards for Tanks—Level 1
PP	• National Emission Standards for Containers
QQ	• National Emission Standards for Surface Impoundments
RR	• National Emission Standards for Individual Drain Systems
SS–UU	• [Reserved]
VV	• National Emission Standards for Oil-Water Separators and Organic-Water Separators
EEE	• National Emission Standards for Hazardous Air Pollutants From Hazardous Waste Combustors
GGG	• National Emission Standards for Pharmaceuticals Production
III	• National Emission Standards for Hazardous Air Pollutants for Flexible Polyurethane Foam Production
JJJ	• National Emission Standards for Hazardous Air Pollutant Emissions: Group IV Polymers and Resins
Part 64	Compliance Assurance Monitoring
Part 66	Assessment and Collection of Noncompliance Penalties by EPA
A	• Purpose and Scope
B	• Notice of Noncompliance
C	• Calculation of Noncompliance Penalties
D	• Exemption Requests; Revocation of Exemptions
E	• Decisions on Exemption Requests and Challenges to Notices of Noncompliance
F	• Review of Penalty Calculation
G	• Payment
H	• Compliance and Final Adjustment
I	• Final Action
J	• Supplemental Rules for Formal Adjudicatory Hearings
Part 67	EPA Approval of State Noncompliance Penalty Program
A	• Purpose and Scope
B	• Approval of State Programs
C	• Federal Notice of Noncompliance to Sources in States With Approved Programs

TABLE 2-1
Master Table of Contents for CAA-Related *Code of Federal Regulations* Entries—Continued

Part/Subpart	*Title*
Part 67—Continued	
D	• EPA Review of State Exemption Decisions
E	• EPA Review of State Penalty Assessments
Part 68	Chemical Accident Prevention Provisions
A	• General
B	• Hazard Assessment
C	• Program 2 Prevention Program
D	• Program 3 Prevention Program
E	• Emergency Response
F	• Regulated Substances for Accidental Release Prevention
G	• Risk Management Plan
H	• Other Requirements
Part 69	Special Exemptions From Requirements of The Clean Air Act
A	• Guam
B	• American Samoa
C	• Commonwealth of the Northern Mariana Islands
Part 70	State Operating Permit Programs
Part 71	Federal Operating Permit Programs
A	• Operating Permits
B	• Permits for Early Reductions Sources
Part 72	Permits Regulation
A	• Acid Rain Program General Provisions
B	• Designated Representative
C	• Acid Rain Permit Applications
D	• Acid Rain Compliance Plan and Compliance Options
E	• Acid Rain Permit Contents
F	• Federal Acid Rain Permit Issuance Procedures
G	• Acid Rain Phase II Implementation
H	• Permit Revisions
I	• Compliance Certification
Part 73	Sulfur Dioxide Allowance System
A	• Background and Summary
B	• Allowance Allocations
C	• Allowance Tracking System
D	• Allowance Transfers
E	• Auctions, Direct Sales, and Independent Power Producers Written Guarantee
F	• Energy Conservation and Renewable Energy Reserve
G	• Small Diesel Refineries

TABLE 2-1

Master Table of Contents for CAA-Related *Code of Federal Regulations* Entries—Continued

Part/Subpart	*Title*
Part 74	Sulfur Dioxide Opt-Ins
A	• Background and Summary
B	• Permitting Procedures
C	• Allowance Calculation for Combustion Sources
D	• Allowance Calculation for Process Sources [Reserved]
E	• Allowance Tracking and Transfer and End of Year Compliance
F	• Monitoring Emissions: Combustion Sources
G	• Monitoring Emissions: Process Sources [Reserved]
Part 75	Continuous Emission Monitoring
A	• General
B	• Monitoring Provisions
C	• Operation and Maintenance Requirements
D	• Missing Data Substitution Procedures
E	• Alternative Monitoring Systems
F	• Recordkeeping Requirements
G	• Reporting Requirements
Part 76	Acid Rain Nitrogen Oxides Emission Reduction Program
Part 77	Excess Emissions
Part 78	Appeal Procedures for Acid Rain Program
Part 79	Registration of Fuels and Fuel Additives
A	• General Provisions
B	• Fuel Registration Procedures
C	• Additive Registration Procedures
D	• Designation of Fuels and Additives
E	• [Reserved]
F	• Testing Requirements for Registration
Part 80	Regulation of Fuels and Fuel Additives
A	• General Provisions
B	• Controls and Prohibitions
C	• Oxygenated Gasoline
D	• Reformulated Gasoline
E	• Anti-Dumping
F	• Attest Engagements
G	• Detergent Gasoline
Part 81	Designation of Areas for Air Quality Planning Purposes
A	• Meaning of Terms

TABLE 2-1
Master Table of Contents for CAA-Related *Code of Federal Regulations* Entries—Continued

Part/Subpart	*Title*
Part 81—Continued	
B	• Designation of Air Quality Control Regions
C	• Section 107 Attainment Status Designations
D	• Identification of Mandatory Class I Federal Areas Where Visibility Is an Important Value
Part 82	Protection of Stratospheric Ozone
A	• Production and Consumption Controls
B	• Servicing of Motor Vehicle Air Conditioners
C	• Ban on Nonessential Products Containing Class I Substances and Ban on Nonessential Products Containing or Manufactured With Class II Substances
D	• Federal Procurement
E	• The Labeling of Products Using Ozone-Depleting Substances
F	• Recycling and Emissions Reduction
G	• Significant New Alternatives Policy Program
H	• Halon Emissions Reduction
Part 88	Clean-Fuel Vehicles
A	• Emission Standards for Clean-Fuel Vehicles
B	• California Pilot Test Program
C	• Clean-Fuel Fleet Program
Part 93	Determining Conformity of Federal Actions to State or Federal Implementation Plans
A	• Conformity to State or Federal Implementation Plans of Transportation Plans, Programs, and Projects Developed, Funded or Approved Under Title 23 U.S.C. or the Federal Transit Act
B	• Determining Conformity of General Federal Actions to State or Federal Implementation Plans

CO = carbon monoxide; NO_2 = nitrogen dioxide; NSO = nonferrous smelter order; O_3 = ozone; PM-10 = particulate matter with an aerodynamic diameter less than or equal to 10 microns; SIP = state implementation plan; SO_2 = sulfur dioxide; U.S.C. = United States Code; VOC = volatile organic compound.

[1]Within this range, Subparts D, G, M, BB, DD, JJ, QQ, and ZZ are reserved because the corresponding states do not have an associated EPA-approved Part 62 plan.

[2]The following states do not have an approved Part 62 plan: Arizona, Colorado, Hawaii, Montana, Nevada, North Dakota, South Dakota, and Wyoming.

Source: Elsevier Science Inc.

TABLE 2-2
Schedule of Upcoming CAA Regulations

Item	CFR Part	Proposed actions: Scheduled date	Proposed actions: Deadline	Proposed actions: APC review[1]	Final actions: Scheduled date	Final actions: Deadline	Final actions: APC review[1]
Rules Under Consideration But Not Formally Announced							
Consolidated emission reporting for state agencies	51	12/98			NYD		
Class II ODS allowance system	82	01/99			11/99		
Rules Announced But Not Yet Proposed							
General conformity rule revisions	51 93	04/98			12/98		
NSR changes to implement revised ozone and particulate matter NAAQS	51 52	04/98			12/98		
Revision to air quality modeling guidelines	51 52	07/98			12/98		
PSD permit application review procedures for nonfederal Class I areas	51 52	**05/16/97** (ANPRM) (62 *FR* 27158)		7:5, 2.12	NA		
		10/98			10/99		
NSR and PSD permitting program for sources in Indian country	51 52	10/98			11/99		
Expansion of source categories whose fugitive emissions must be considered for PSD and nonattainment NSR applicability	51 52 70 71	02/99			02/00		
FIP for VOC reductions in Washington, D.C.	52	06/98	05/98 J		01/99	01/96 S 12/98 J	
Controls to address regional NO_x emissions transported to northeast states (based on CAA Section 126 petition)	52	09/98			04/99	04/98 S	
FIP for regional NO_x transport control in Eastern states	52	10/98			12/99	10/99 S	
VOC Control Regulations for Consumer and Commercial Products							
—Offset lithographic printing	59	04/98			03/99		
Consolidated emission standard for SOCMI facilities	60 61 63	04/98			NYD		

TABLE 2-2
Schedule of Upcoming CAA Regulations—Continued

Item	CFR Part	Proposed actions			Final actions		
		Scheduled date	Deadline	APC review[1]	Scheduled date	Deadline	APC review[1]
Rules Announced But Not Yet Proposed—Continued							
Revised stack flow measurement rules	60 75	06/98			03/99		
FIPs for two power stations on Navajo land	60	09/98			NYD		
Predictive emission monitoring system guidelines	60	09/98			NYD		
Amendment to water-based coating test (Method 24)	60	11/98			10/99		
Correction and clarification of air pollution analysis procedures (Method 301)	60 63	12/98			01/99		
Storage tank provision revisions	60 61 63 264 265	05/99			NYD		
NSPS Provisions							
—Revised rule for electric arc furnaces	60	NA[2]			06/98		
—Sewage sludge incinerators	60	05/99			05/00	11/00 S	
—Combustion turbine standards	60	11/99			11/00	11/00 S	
—Industrial and commercial waste incinerators	60	11/99			11/00	11/94 S	
—Other solid waste incinerators	60	11/99			11/00		
—Residential wood heaters	NYD	NA			03/99 (Existing rule review)		
General NESHAP Provisions							
—Amendments to general NESHAP provisions	63	04/98			03/99		
—Regulations for approval of state air toxics programs in lieu of MACT standards	63	05/98			05/99		
—Potential-to-emit determination guidelines	63	08/98			08/99		

TABLE 2-2
Schedule of Upcoming CAA Regulations—Continued

Item	CFR Part	Proposed actions: Scheduled date	Proposed actions: Deadline	Proposed actions: APC review[1]	Final actions: Scheduled date	Final actions: Deadline	Final actions: APC review[1]
Rules Announced But Not Yet Proposed—Continued							
General NESHAP Provisions—Continued							
—Addition of cyanide chemical manufacturers to source categories regulated under the HON rule	63	11/99			11/00	11/00 S	
MACT Standards							
—Automobile and light-duty truck manufacturing[3]	59 63	11/99			11/00	11/00 S	
—Flatwood paneling coating[3]	59 63	11/99			11/00	11/00 S	
—Large appliance coatings[3]	59 63	11/99			11/00	11/00 S	
—Metal furniture coatings[3]	59 63	11/99			11/00	11/00 S	
—Miscellaneous metal parts and products coatings[3]	59 63	11/99			11/00	11/00 S	
—Paper, film, and foil coating[3]	59 63	11/99			11/00	11/00 S	
—Plastic part coatings[3]	59 63	11/99			11/00	11/00 S	
—Offsite waste recovery operations standard revisions	63	04/98			06/98		
—Publicly owned treatment works	63	04/98			03/99	11/95 S	
—Revisions to chromium electroplating standards to address advanced process sources	63	NA[2]			04/98		
—Alumina processing	63	04/98			NYD	11/00 S	
—Revised standards for continuous web cleaners from halogenated solvent cleaning MACT	63	NA			**05/05/98** (Three-month stay) (63 *FR* 24749)		8:5, 2.52
		05/05/98 (One-year stay) (63 *FR* 24765)		8:5, 2.52	NYD		
		04/98			NYD		
—HON rule revisions	63	06/98			12/98		

TABLE 2-2
Schedule of Upcoming CAA Regulations—Continued

Item	CFR Part	Proposed actions			Final actions		
		Scheduled date	Deadline	APC review[1]	Scheduled date	Deadline	APC review[1]
Rules Announced But Not Yet Proposed—Continued							
MACT Standards—Continued							
—Revision of Groups I and IV polymer and resin standards	63	NA			**03/31/98** (Compliance extension) (63 *FR* 15312)[4]		8:4, 2.52
		06/98			04/99		
—Revisions to magnetic tape manufacturing standards	63	07/98			12/98		
—Wet-formed fiberglass mat production	63	07/98			07/99	11/00 S	
—Widening of applicability of chromium electroplating standards	63	08/98			02/99		
—Secondary aluminum refiners	63	09/98			04/99	11/97 S	
—Asphalt roofing and processing	63	10/98			10/99	11/00 S	
—Ethylene processes	63	11/98			11/99	11/00 S	
—Flexible polyurethane foam fabrication	63	02/99			02/00	11/00 S	
—Coke ovens—pushing, quenching, and battery stacks	63	05/99			05/00	11/00 S	
—Primary magnesium refining	63	05/99			05/00	11/00 S	
—Chromium refractories	63	05/99			11/00	11/00 S	
—Friction products	63	05/99			11/00	11/00 S	
—Polyvinyl chloride production	63	06/99			06/00	11/00 S	
—Semiconductor manufacturing	63	07/99			09/00	11/00 S	
—Organic liquid distribution	63	07/99			11/00	11/00 S	
—Asphalt/coal tar application on metal pipes	63	09/99			01/01	11/00 S	
—Clay products manufacturing	63	09/99			01/01	11/00 S	
—Carbon black production	63	11/99[5]			11/00	11/00 S	
—Chlorine production	63	11/99			11/00	11/00 S	
—Contaminated site remediation	63	11/99			11/00	11/00 S	
—Fabric printing, coating, and dyeing	63	11/99			11/00	11/00 S	
—Fumed silica production	63	11/99			11/00	11/00 S	

©Elsevier Science Inc. 0-444-10032-6/98/$19.00+$5.50

TABLE 2-2
Schedule of Upcoming CAA Regulations—Continued

Item	CFR Part	Proposed actions			Final actions		
		Scheduled date	Deadline	APC review[1]	Scheduled date	Deadline	APC review[1]
Rules Announced But Not Yet Proposed—Continued							
MACT Standards—Continued							
—Industrial, commercial, and institutional boilers and process heaters	63	11/99			11/00	11/00 S	
—Integrated iron and steel mills	63	11/99			11/00	11/00 S	
—Iron and steel foundries	63	11/99			11/00	11/00 S	
—Metal can coating	63	11/99			11/00	11/00 S	
—Metal coil coating	63	11/99			11/00	11/00 S	
—Miscellaneous organic chemical production	63	11/99			11/00	11/97 S	
—Paint stripping operations	63	11/99			11/00	11/00 S	
—Plywood and particle board manufacturing	63	11/99			11/00	11/00 S	
—Reinforced composite plastic producers	63	11/99			11/00	11/00 S	
—Rocket engine test firing facilities	63	11/99			11/00	11/00 S	
—Spandex production	63	11/99			11/00	11/00 S	
—Tire manufacturing	63	11/99			11/00	11/00 S	
—Lime manufacturing	63	11/99			NYD	11/00 S	
—Municipal solid waste landfills	63	11/99			NYD	11/00 S	
—Petroleum solvent dry cleaners	63	11/99			NYD	11/00 S	
—Ammonium sulfate production—caprolactam by-product plants	63	12/99			12/00	11/00 S	
—Boat manufacturing	63	12/99			12/00	11/00 S	
—Uranium hexafluoride production	63	12/99			12/00	11/00 S	
—Cellulose production	63	12/99			NYD	11/00 S	
—Hydrogen chloride production	63	11/00			11/01	11/00 S	
—Hydrochloric acid production	63	11/00			NYD	11/00 S	
—Leather tanning and finishing operations	63	NYD			NYD	11/00 S	
—Coordinated rulemaking for industrial combustion units	NYD	11/99			11/00	11/00 S	
—Vegetable oil production	NYD	11/99			11/00	11/00 S	
—Taconite iron ore processing	NYD	12/01			12/02	11/00 S	

TABLE 2-2
Schedule of Upcoming CAA Regulations—Continued

Item	CFR Part	Proposed actions: Scheduled date	Proposed actions: Deadline	Proposed actions: APC review[1]	Final actions: Scheduled date	Final actions: Deadline	Final actions: APC review[1]
Rules Announced But Not Yet Proposed—Continued							
Clarified applicable requirement provisions for operating permit applications	70	08/98			06/99		
Review of operating permits issued by Indian tribes	70	11/98			03/99		
Modified permit revision procedure for federal operating permit program	71	12/98			07/99		
Widening of ban on sale of certain ODS for nonessential uses	82	06/98			01/99		
Reconsideration of sales ban on precharged split-systems	82	07/98			01/99		
Revised standards for MVAC refrigerant recovery/recycle equipment	82	08/98			01/99		
Supplemental labeling rule for HCFCs	82	12/98			06/99		
Revised pollution standard index and significant harm level for particulate and ozone	NYD	04/98			12/98		
FIP for the Fort Hall Indian Reservation	NYD	04/98			NYD		
Rules Proposed But Not Yet Finalized							
Review of primary NAAQS for SO_x	50 51	**11/15/94** (Review decision) (59 *FR* 58958)	11/94 J	5:2, 2.35	NA		
		03/07/95 (Review decision and implementation plan) (60 *FR* 12492)		5:3, 2.18	**05/22/96** (61 *FR* 25566)	04/96 J	6:6, 2.42
		01/02/97 (SO_x intervention plan) (62 *FR* 210)		7:2, 2.33	12/00[6]		

TABLE 2-2
Schedule of Upcoming CAA Regulations—Continued

Item	CFR Part	Proposed actions: Scheduled date	Proposed actions: Deadline	Proposed actions: APC review[1]	Final actions: Scheduled date	Final actions: Deadline	Final actions: APC review[1]
Rules Proposed But Not Yet Finalized—Continued							
Review of primary NAAQS for SO_x—Continued							
		05/05/98 (Action plan notice) (63 *FR* 24782)		8:5, 2.54	NA		
		09/99 (Revised review decision)[6]			12/99[6]		
Regulatory burden reduction	51 60 61 63	**09/11/96** (61 *FR* 47840)		7:1, 2.40	04/98		
Requirements for ozone and particulate matter SIPs and regional haze control	51 81	**12/13/96** (Key issues ANPRM) (61 *FR* 65764)		7:2, 4.1	NA		
		12/13/96 (Implementation policy notice) (61 *FR* 65752)		7:2, 4.1	NA		
		NA			**01/16/98** (1-hr Ozone NAAQS exemptions) (63 *FR* 2726)[4]		8:3, 2.2
		NA			**03/16/98** (Withdrawal of 1-hr Ozone NAAQS exemptions) (63 *FR* 12652)		8:4, 2.47
		NA			**07/22/98** (Additional 1-hr Ozone NAAQS exemptions) (63 *FR* 39432)		8:6, 2.25

TABLE 2-2
Schedule of Upcoming CAA Regulations—Continued

Item	*CFR Part*	*Proposed actions*			*Final actions*		
		Scheduled date	*Deadline*	*APC review*[1]	*Scheduled date*	*Deadline*	*APC review*[1]
Rules Proposed But Not Yet Finalized—Continued							
Requirements for ozone and particulate matter SIPs and regional haze control—Continued							
		01/16/98 (1-hr Ozone NAAQS exemptions) (63 *FR* 2804)		8:3, 2.2	**06/05/98** (63 *FR* 31014)		8:5, 2.6
		07/31/97 (Regional haze plan) (62 *FR* 41138)		7:6, 2.31	04/98		
		04/98 (Antiback-sliding rule for existing PM-10 NAAQS)			09/98		
		04/98 (Conformity rule for tran-sitional areas)			12/98		
		06/98 (Implementa-tion planning guidance)			12/98		
Open market trading of ozone pre-cursor emissions	51	**08/03/95** (Policy statement) (60 *FR* 39668)		5:6, 2.26	05/98		
		08/25/95 (Correction notice) (60 *FR* 44290)		5:6, 2.26	NA		
Test Methods 203A–203C for vis-ible emission measurement	51	**11/22/93** (58 *FR* 61640)		4:2, 2.42	08/98		

TABLE 2-2
Schedule of Upcoming CAA Regulations—Continued

Item	*CFR Part*	*Proposed actions*			*Final actions*		
		Scheduled date	*Deadline*	*APC review*[1]	*Scheduled date*	*Deadline*	*APC review*[1]
Rules Proposed But Not Yet Finalized—Continued							
SIP call for emission controls on regionally transported ozone precursors	51 52	**11/07/97** (Proposed NO_x budgets) (62 *FR* 60318)		8:2, 2.2	NA		
		05/11/98 (Regulatory text) (63 *FR* 25902)		8:5, 2.12	09/98		
Opacity test method for stationary sources	51	**10/07/92** (57 *FR* 46114)		3:1, 2.34	10/98		
NSR reform	51 52	**07/23/96** (61 *FR* 38250)		6:4, 4.1	12/98		
		07/24/98 (Notice) (63 *FR* 39857)		8:6, 2.26	NA		
Test Method 207 for airborne isocyanate compounds	51	**12/08/97** (63 *FR* 64532)		8:2, 2.41	12/98		
Revisions to operating permit program to address litigation issues and other problems	51 70 71	**08/29/94** (59 *FR* 44460)		4:7, 2.1	NA		
		08/31/95 (Supplemental proposal) (60 *FR* 45530)		5:6, 2.1	04/99		
Elimination of selected lead ambient air monitoring stations[7]	58	NA			**11/05/97** (62 *FR* 59813)[4]		8:2, 2.40
		NA			**12/23/97** (Withdrawal) (63 *FR* 67009)		8:2, 2.40
		11/05/97 (62 *FR* 59840)		8:2, 2.40	NYD		
Field citation enforcement program	59	**05/03/94** (59 *FR* 22776)		4:4, 2.22	04/98		
VOC Control Regulations for Consumer and Commercial Products							
—Aerospace, shipbuilding, and wood furniture coatings	59	**08/22/97** (62 *FR* 44672)[8]		7:6, 2.41	12/98	03/97 S	

TABLE 2-2
Schedule of Upcoming CAA Regulations—Continued

Item	CFR Part	Proposed actions: Scheduled date	Proposed actions: Deadline	Proposed actions: APC review[1]	Final actions: Scheduled date	Final actions: Deadline	Final actions: APC review[1]
Rules Proposed But Not Yet Finalized—Continued							
VOC Control Regulations for Consumer and Commercial Products—Continued							
—Standardization of emission test methods	60 61 63	**08/27/97** (62 *FR* 45369)		7:6, 2.45	09/98		
		01/14/98 (Supplemental proposal) (63 *FR* 2194)		8:3, 2.32	NA		
NSPS Provisions							
—Revision of opacity monitoring provisions	60	**11/25/94** (59 *FR* 60585)		5:2, 2.36	NA		
		09/23/98 (Supplemental proposal) (63 *FR* 50824)		9:1[9]	03/99		
—SOCMI wastewater facilities	60	**09/12/94** (59 *FR* 46780)	08/94 J	5:1, 2.1	NA		
		10/11/95 (Supplemental proposal) (60 *FR* 52889)		6:1, 2.30	NA		
		04/98 (Second supplemental proposal)			04/00		
Radon emission allowance for research use of phosphogypsum	61	**03/24/94** (Notice) (59 *FR* 14040)		4:4, 2.56	NA		
		05/08/96 (61 *FR* 20775)		6:6, 2.36	04/98		
General NESHAP Provisions							
—National urban air toxics strategy	63	**09/14/98** (Draft strategy) (63 *FR* 49240)	08/98 J	9:1[9]	NYD	11/95 S 06/99 J	

©Elsevier Science Inc. 0-444-10032-6/98/$19.00+$5.50

TABLE 2-2
Schedule of Upcoming CAA Regulations—Continued

Item	*CFR Part*	*Proposed actions*			*Final actions*		
		Scheduled date	*Deadline*	*APC review*[1]	*Scheduled date*	*Deadline*	*APC review*[1]
Rules Proposed But Not Yet Finalized—Continued							
MACT Standards							
—Phosphoric acid manufacturing and phosphate fertilizer production	63	**12/27/96** (61 *FR* 68430)		7:2. 2.14	04/98	11/97 S	
—Steel pickling—hydrochloric acid process	63	**09/18/97** (62 *FR* 49052)		8:1, 2.27	04/98	11/97 S	
—Wool fiberglass manufacturing industry	63	**03/31/97** (62 *FR* 15228)		7:4, 2.16	04/98	11/97 S	
—Mineral wool production industry	63	**05/08/97** (62 *FR* 25370)		7:4, 2.40	05/98	11/97 S	
—Feroalloy industry	63	**08/04/98** (63 *FR* 41508)		8:6, 2.30	09/98	11/97 S	
—Primary copper smelters	63	**04/20/98** (63 *FR* 19582)		8:4, 2.34	09/98	11/97 S	
—Primary lead smelters	63	**04/17/98** (63 *FR* 19200)		8:4, 2.43	09/98	11/97 S	
—Polyether polyol production	63	**09/04/97** (62 *FR* 46804)		7:6, 2.10	11/98	11/97 S	
—MACT standards for hazardous waste combustion facilities	60 63 260 261 264 265 266 270 271	**04/19/96** (Industrial furnaces and incinerators) (61 *FR* 17358)		6:4, 2.2	12/98	04/98 J	
		NA			**06/19/98** ("Fast track" rule) (63 *FR* 33782)		8:5, 2.16
		12/99 (Boilers)			NYD	12/99 J	
—Revisions to wood furniture standards[10]	63	**06/24/98** (63 *FR* 34336)	06/98 J	8:6. 2.11	01/99	12/98 J	
—Portland cement manufacturers	63	**03/24/98** (63 *FR* 14182)		8:4, 2.13	02/99	11/97 S	
—Agricultural chemical producers	63	**11/10/97** (62 *FR* 60566)		8:2, 2.6	03/99	11/97 S	

TABLE 2-2
Schedule of Upcoming CAA Regulations—Continued

Item	CFR Part	Proposed actions: Scheduled date	Proposed actions: Deadline	Proposed actions: APC review[1]	Final actions: Scheduled date	Final actions: Deadline	Final actions: APC review[1]
Rules Proposed But Not Yet Finalized—Continued							
MACT Standards—Continued							
—Miscellaneous source category standards, including production of acrylic/modacrylic fibers, formaldehyde-based polymers and resins, hydrogen fluoride, and polycarbonate resins	63	**10/14/98** (63 *FR* 55177)		9:2[11]	05/99	11/97 S	
—Oil and natural gas production facilities	63	**02/06/98** (63 *FR* 6288)		8:3, 4.1	05/99	11/97 S	
—Petroleum refinery fluidized catalytic cracking units, reformers, and sulfur plants	63	**09/11/98** (63 *FR* 48890)		9:1[9]	05/99	11/97 S	
—Baker's yeast manufacturing	63	**10/19/98** (63 *FR* 55812)		9:2[11]	09/99	11/00 S	
Awards for information leading to CAA violation convictions	65	**05/03/94** (59 *FR* 22795)		4:4, 2.24	NYD		
Accidental release prevention program revisions	68	**04/17/98** (63 *FR* 19216)		8:4, 2.22	06/98		
Federal operating permit program for Indian lands	71	**03/21/97** (62 *FR* 13748)		7:4, 2.22	05/98	11/97 S	
SO_2 allowance system revisions	72 73	**08/03/98** (63 *FR* 41358)		8:6, 2.23	10/98		
Acid rain program continuous emission monitoring	72 75	**05/21/98** (63 *FR* 28032)		8:5, 2.23	01/99		
Deletion of specific units from acid rain program and adjustment of SO_2 allowance allocations to others	73	**12/27/96** (61 *FR* 68340)		7:2, 2.20	04/98		
Results of agency analysis of adequacy of bias test results, relative accuracy requirements, and missing data substitution procedure triggers	75	**05/21/98** (63 *FR* 28195)		8:5, 2.23	NYD		
Updated refrigerant reclaiming/recycling standard	82	**02/29/96** (61 *FR* 7858)		6:3, 2.21	NA		
		06/11/98 (Reproposal) (63 *FR* 32044)		8:5, 2.2	04/98		

TABLE 2-2
Schedule of Upcoming CAA Regulations—Continued

Item	*CFR Part*	*Proposed actions*			*Final actions*		
		Scheduled date	*Deadline*	*APC review*[1]	*Scheduled date*	*Deadline*	*APC review*[1]
Rules Proposed But Not Yet Finalized—Continued							
Inclusion of all refrigerants in recycling rule	82	**06/11/98** (63 *FR* 32044)		8:5, 2.2	10/98	11/95 S	
Reporting requirements for ODS import petition process	82	NA			**08/4/98** (63 *FR* 41626)[4]		8:6, 2.2
		NA			**10/05/98** (Withdrawal) (63 *FR* 53290)		9:1[9]
		08/04/98 (63 *FR* 41652)		8:6, 2.2	NYD		
Amendment of hazardous waste treatment, storage, and disposal facility air emission standards	264 265 270	NA			**02/09/96** (Clarifying amendments) (61 *FR* 4903)		6:3, 2.36
		NA			**11/25/96** (Revisions and effective date postponement) (61 *FR* 59932)		7:2, 2.34
		NA			**12/08/97** (Technical correction) (62 *FR* 64636)		8:2, 2.25
		NA			05/98 (Additional technical changes)		
Rules That Have Been Finalized Since April 27, 1998							
VOC Control Regulations for Consumer and Commercial Products							
—Architectural coatings	59	**06/25/96** (61 *FR* 32729)		6:6, 2.29	**09/11/98** (63 *FR* 48806)	03/97 S	9.1[9]
—Auto refinish coatings	59	**04/30/96** (61 *FR* 19005)		6:4, 2.23	**09/11/98** (63 *FR* 48806)	03/97 S	9:1[9]
		12/30/97 (Supplemental proposal) (62 *FR* 67784)		8:2, 2.32	NA		

TABLE 2-2
Schedule of Upcoming CAA Regulations—Continued

Item	CFR Part	Proposed actions: Scheduled date	Proposed actions: Deadline	Proposed actions: APC review[1]	Final actions: Scheduled date	Final actions: Deadline	Final actions: APC review[1]
Rules That Have Been Finalized Since April 27, 1998—Continued							
VOC Control Regulations for Consumer and Commercial Products—Continued							
—Consumer products	59	**04/02/96** (61 *FR* 14531)		6:4, 2.23	**09/11/98** (63 *FR* 48806)	03/97 S	9.1[9]
Hydrogen-fueled flare general NSPS and NESHAP rules	60 63	NA			**05/04/98** (63 *FR* 24436)[4]		8:5, 2.53
NSPS Provisions							
—Revised standards for NO_x emissions from steam-generating units	60	**07/09/97** (62 *FR* 36948)	07/97 J	7:5, 2.5	**09/16/98** (63 *FR* 49442)	09/98 J	9:1[9]
MACT Standards							
—Tetrahydrobenzaldehyde (butadiene dimers) producers	63	**08/22/97** (62 *FR* 44614)		7:6, 2.43	**05/12/98** (63 *FR* 26078)[12]	11/97 S	8:5, 2.19
—Petroleum refinery standard revisions	63	NA			**03/20/98** (63 *FR* 13533)[4]		8:4, 2.49
		NA			**05/18/98** (Withdrawal) (63 *FR* 27212)		8:5, 2.51
		03/20/98 (63 *FR* 13587)		8:4, 2.49	**08/18/98** (63 *FR* 44135)		8:6, 2.16
		NA			**06/09/98** (Implementation plan deadline, et al.) (63 *FR* 31358)[4]		8:5, 2.56
—Pharmaceutical manufacturers	63	**04/02/97** (62 *FR* 15754)		7:4, 2.2	**09/21/98** (63 *FR* 50280)	11/97 S	9:1[9]
—Flexible polyurethane foam production	63	**12/27/96** (61 *FR* 68406)		7:2, 2.25	**10/07/98** (63 *FR* 53980)	11/97 S	9:1[9]
Revision of acid rain SO_2 allowance allocations	73	**01/7/98** (63 *FR* 714)		8:3, 2.29	**09/28/98** (63 *FR* 51706)	06/98 S	9:1[9]

TABLE 2-2
Schedule of Upcoming CAA Regulations—Continued

Item	CFR Part	Proposed actions			Final actions		
		Scheduled date	Deadline	APC review[1]	Scheduled date	Deadline	APC review[1]
Clarification of ODS labeling requirements to fresh foods treated with methyl bromide	82	NA			**05/15/98** (Notice) (63 *FR* 26983)	05/98 J	8:5, 2.47

ANPRM = advanced notice of proposed rulemaking; *APC* = *The Air Pollution Consultant*; BAC = best available control; CTG = control techniques guideline; FIP = federal implementation plan; HAP = hazardous air pollutant; HCFC = hydrochlorofluorocarbon; HON = hazardous organic NESHAP; MACT = maximum achievable control technology; MVAC = motor vehicle air conditioner; NA = not applicable; NAAQS = national ambient air quality standards; NESHAP = national emission standards for hazardous air pollutants; NO_x = nitrogen oxides; NSPS = new source performance standards; NSR = new source review; NYD = not yet determined; ODS = ozone-depleting substances; PSD = prevention of significant deterioration; SIP = state implementation plan; SOCMI = synthetic organic chemical manufacturing industry; SO_2 = sulfur dioxide; SO_x = sulfur oxides; VOC = volatile organic compounds.

[1]Refers to the volume, issue, and page number of the applicable *APC* review article. For example, 7:5, 2.12 refers to the *APC* article beginning on page 2.12 of Volume 7, Issue 5.

[2]A direct final rule is planned.

[3]This rulemaking will contain both MACT standards for HAP control and BAC standards for VOC control.

[4]Issued as a direct final rule.

[5]This item is listed twice in the regulatory agenda. One listing gives this date as 08/99, the other lists 11/99. The later date is probably more realistic.

[6]Date listed is from May 5, 1998 notice (see 63 *FR* 24783).

[7]According to regulatory agenda, EPA has combined this action with another item and withdrawn this entry. However, this second entry is not listed in the *Federal Register*; in addition, other sources indicate that final action on this entry *is* still planned. Thus, this entry has been left in our listing.

[8]EPA has proposed that CTGs issued for each of the three subject industries provide sufficient control and, therefore, preclude the need for issuance of a Part 59 control regulation.

[9]This action will be reviewed in the January/February 1999 issue of *APC*; the page on which the article will appear is not yet known.

[10]This item is listed twice in the regulatory agenda; where conflicting data are presented, dates listed are those from the newer entry.

[11]This action will be reviewed in the March/April 1999 issue of *APC*; the page on which the article will appear is not yet known.

[12]EPA has imposed MACT on this category of sources by adding tetrahydrobenzaldehyde to the list of chemicals regulated under the HON rule; in the listed final action, the agency also deleted tetrahydrobenzaldehyde production from the list of source categories to be regulated under a separate MACT standard.

Source: Compiled by Elsevier Science Inc.

3 The National Ambient Air Quality Standards and New Source Review Programs

Title I of the CAA spells out, among other things, the national ambient air quality standards (NAAQS) and new source review (NSR) programs (including the prevention of significant deterioration [PSD] program and nonattainment NSR program). These sections of the statute have evolved significantly over the last 25 years. Therefore, as shown in Table 3-1, the corresponding regulations are arranged in a piecemeal fashion, making them rather difficult to follow.

NAAQS

The NAAQS program regulates six criteria pollutants: ozone, carbon monoxide (CO), particulate matter (regulated as PM-2.5 and PM-10 [particulate matter with an aerodynamic diameter less than or equal to 2.5 microns and 10 microns, respectively]), sulfur oxides or SO_x (which are measured as sulfur dioxide [SO_2]), nitrogen dioxide (NO_2), and lead. The current NAAQS, including the landmark revisions to the ozone and particulate matter NAAQS that were promulgated on July 18, 1997 (62 *FR* 38652 and 38856), are listed in Table 3-2 (page 3.2). As shown in the table, two types of NAAQS are required for each of the criteria pollutants (with the exception of CO): 1) *primary standards* designed to protect public health, and 2) *secondary standards* designed to protect public welfare including economic interests, vegetation, visibility, and other factors.

The NAAQS represent maximum acceptable concentrations of criteria pollutants in the atmosphere. From a national perspective, they establish uniform targets against which the ambient air quality of all areas of the country are measured.

Chemicals Excluded From the Definition of "VOCs"

As mentioned above, ozone is regulated as a criteria pollutant under the NAAQS program. However, it is a "secondary

TABLE 3-1

Location of Select Regulations Governing NAAQS, NSR, and PSD

Regulation	Subject
40 *CFR* Part 50	National primary and secondary ambient air quality standards
§§50.4–50.12	—National ambient air quality standards (NAAQS)
Appendices A–J	—Reference methods for measurement of pollutants in the atmosphere and interpretation of NAAQS
40 *CFR* Part 51	Requirements for preparation, adoption, and submittal of implementation plans
§51.165	—Permit requirements (nonattainment NSR)
§51.166	—Prevention of significant deterioration of air quality (PSD)
40 *CFR* Part 52	Approval and promulgation of implementation plans
§52.21	—Prevention of significant deterioration of air quality
§52.24	—Statutory restriction on new sources (nonattainment NSR)
§§52.50–52.2920	—State-specific provisions
40 *CFR* Part 53	Ambient air monitoring reference and equivalent methods
40 *CFR* Part 55	Outer continental shelf air regulations
40 *CFR* Part 56	Regional consistency
40 *CFR* Part 57	Primary nonferrous smelter orders
40 *CFR* Part 58	Ambient air quality surveillance
40 *CFR* Part 59	National volatile organic compound emission standards for consumer and commercial products
40 *CFR* Part 81	Designation of areas for air quality planning purposes

NAAQS = national ambient air quality standards, NSR = new source review; PSD = prevention of significant deterioration.

Source: Elsevier Science Inc.; compiled from 40 *CFR* Parts 50–58, and Part 81.

TABLE 3-2
National Ambient Air Quality Standards

Criteria pollutant	*Standard type*	*NAAQS limit*		*Compliance determination basis*[a]
		μg/m³	*ppm*	
Ozone	Primary and secondary	—	0.08	3-year average of fourth highest daily maximum 8-hour average concentrations
	Primary and secondary	235[b]	0.12[b]	Maximum hourly average concentration; one exceedance allowed per year
CO	Primary	10,000	9	8-hour average concentration; one exceedance allowed per year
	Primary	40,000	35	1-hour average concentration; one exceedance allowed per year
PM-2.5	Primary and secondary	65	—	3-year average of the 98th percentile 24-hour average concentrations
	Primary and secondary	15	—	3-year average of annual arithmetic mean concentrations
PM-10	Primary and secondary	150	—	3-year average of the 99th percentile 24-hour average concentrations
	Primary and secondary	150	—	Maximum 24-hour average concentration; one exceedance allowed per year[c,d]
	Primary and secondary	50	—	3-year average of the annual arithmetic mean concentrations
SO_x [e]	Primary	80	0.03	Annual arithmetic mean
	Primary	365	0.14	Maximum 24-hour concentration; one exceedance allowed per year
	Secondary	1,300	0.5	Maximum 3-hour concentration; one exceedance allowed per year
NO_2	Primary and secondary	100	0.053	Annual arithmetic mean
Lead	Primary and secondary	1.5	—	Maximum arithmetic mean measured over a calendar quarter

CO = carbon monoxide; NAAQS = national ambient air quality standards; NO_2 = nitrogen dioxide; PM-2.5 = particulate matter with an aerodynamic diameter less than or equal to 2.5 microns; PM-10 = particulate matter with an aerodynamic diameter less than or equal to 10 microns; SIP = state implementation plan; SO_x = sulfur oxides; SO_2 = sulfur dioxide.

[a]If the specified value is less than or equal to the listed NAAQS value, the area would be in attainment of the NAAQS.

[b]This limit is applicable only to areas designated nonattainment for ozone in 40 *CFR* Part 81 as of July 18, 1997; once an area has attained the 1-hour NAAQS and has been redesignated accordingly, the 1-hour limit will no longer apply.

[c]Exceedances per year averaged over a 3-year period.

[d]This standard will continue to apply to particulate nonattainment areas until EPA either 1) approves a SIP for the area that includes the needed control measures, or 2) promulgates a rule under CAA Section 172(e) that specifies required PM-10 control measures applicable to that area.

[e]Measured as SO_2.

Source: Elsevier Science Inc.; compiled from 40 *CFR* §§50.4–50.12.

pollutant." In other words, little ozone is actually emitted by sources; rather, ground-level ozone forms as other pollutants react in the lower atmosphere. Thus, to control ozone, EPA regulates the pollutant's precursors: nitrogen oxides (NO_x) and volatile organic compounds (VOCs). (Note the important distinction between *ground-level* ozone, also called smog, and *stratospheric* ozone, a material located in the upper atmosphere that shields the earth from harmful ultraviolet radiation. CAA Title I seeks to reduce ground-level ozone while Title VI seeks to protect stratospheric ozone.)

Most VOCs act as ozone precursors because they are photochemically reactive—in the presence of sunlight, they participate in a series of chemical reactions leading to ozone formation. Consequently, certain nonphotochemically reactive organic compounds are not regulated as VOCs because they do not participate significantly in ozone-forming atmospheric reactions. EPA's list of VOC-exempt chemicals [i.e., those that do not fit the definition of "VOCs" codified at §51.100(s) and therefore are specifically excluded from that definition] appears in Table 3-3. Due to their exemption, these chemicals are not regulated under the nonattainment NSR program or the PSD program. Furthermore, they do not trigger imposition of reasonably available control technology (RACT) requirements for existing sources in NAAQS nonattainment areas.

TABLE 3-3
Chemicals Excluded From the Definition of VOC

Chemical name/formula	*Synonym/alternative description*
Acetone	—
Ethane	—
Methane	—
Methyl acetate	—
Methylene chloride	Dichloromethane
Parachlorobenzotrifluoride	PCBTF
Perchloroethylene	Tetrachloroethylene
1,1,1-Trichloroethane	Methyl chloroform
$(CF_3)_2CFCF_2OCH_3$	2-(Difluoromethoxymethyl)-1,1,1,2,3,3,3-heptafluoropropane
$(CF_3)_2CFCF_2OC_2H_5$	2-(Ethoxydifluoromethyl)-1,1,1,2,3,3,3-heptafluoropropane
$C_4F_9OCH_3$	1,1,1,2,2,3,3,4,4-Nonafluoro-4-methoxybutane
$C_4F_9OC_2H_5$	1-Ethoxy-1,1,2,2,3,3,4,4,4-nonafluorobutane
Cyclic, branched, or linear completely methylated siloxanes	—
Perfluorocarbon compounds in the following classes:	
Cyclic, branched, or linear, completely fluorinated alkanes;	—
Cyclic, branched, or linear, completely fluorinated ethers with no unsaturations;	—
Cyclic, branched, or linear, completely fluorinated tertiary amines with no unsaturations; and	—
Sulfur-containing perfluorocarbons with no unsaturations and with sulfur bonds only to carbon and fluorine.	—

TABLE 3-3
Chemicals Excluded From the Definition of VOC—Continued

Chemical name/formula	*Synonym/alternative description*
CFCs:	
Trichlorofluoromethane	CFC-11
Dichlorodifluoromethane	CFC-12
1,1,2-Trichloro-1,2,2-trifluoroethane	CFC-113
1,2-Dichloro-1,1,2,2-tetrafluoroethane	CFC-114
Chloropentafluoroethane	CFC-115
HCFCs:	
Chlorodifluoromethane	HCFC-22
Chlorofluoromethane	HCFC-31
1,1,1-Trifluoro-2,2-dichloroethane	HCFC-123
1,2-Dichloro-1,1,2-trifluoroethane	HCFC-123a
2-Chloro-1,1,1,2-tetrafluoroethane	HCFC-124
1,1-Dichloro-1-fluoroethane	HCFC-141b
1-Chloro-1,1-difluoroethane	HCFC-142b
1-Chloro-1-fluoroethane	HCFC-151a
3,3-Dichloro-1,1,1,2,2-pentafluoropropane	HCFC-225ca
1,3-Dichloro-1,1,2,2,3-pentafluoropropane	HCFC-225cb
HFCs:	
Trifluoromethane	HFC-23
Difluoromethane	HFC-32
1,1,1,2,3,4,4,5,5,5-Decafluoropentane	HFC-43-10mee
Pentafluoroethane	HFC-125
1,1,2,2-Tetrafluoroethane	HFC-134
1,1,1,2-Tetrafluoroethane	HFC-134a
1,1,1-Trifluoroethane	HFC-143a
1,1-Difluoroethane	HFC-152a
Ethylfluoride	HFC-161
1,1,1,2,3,3-Hexafluoropropane	HFC-236ea
1,1,1,3,3,3-Hexafluoropropane	HFC-236fa
1,1,2,2,3-Pentafluoropropane	HFC-245ca
1,1,2,3,3-Pentafluoropropane	HFC-245ea
1,1,1,2,3-Pentafluoropropane	HFC-245eb
1,1,1,3,3-Pentafluoropropane	HFC-245fa
1,1,1,3,3-Pentafluorobutane	HFC-365mfc

CFC = chlorofluorocarbon; HCFC = hydrochlorofluorocarbon; HFC = hydrofluorocarbon; VOC = volatile organic compound.

Source: Elsevier Science Inc.; compiled from §51.100(s).

Nonattainment Areas

Cities or regions that fail to meet one or more of the NAAQS are designated as nonattainment areas for the pollutant(s) of concern and are listed as such in 40 *CFR* Part 81, Subpart C. In fact, many major urban areas of the United States are designated nonattainment for at least one criteria pollutant.

The 1990 amendments require EPA to classify nonattainment areas for ozone, CO, and PM-10 based on the severity of the observed NAAQS violations in each area. Table 3-4 lists the classifications and compliance deadlines for nonattainment areas. To compel areas to eventually reach full NAAQS attainment, the 1990 amendments require that controls for sources in nonattainment areas become progressively more stringent as the NAAQS violation (and resulting area classification) becomes more severe. In addition and as shown in Table 3-4, as an area's nonattainment classification becomes more severe, the required NAAQS attainment date is also extended.

Under the CAA, many questions of applicability turn on whether a given facility is considered a major source. To make matters more confusing, "major source" is defined differently for attainment versus nonattainment areas. Major

TABLE 3-4
Requirements for NAAQS Nonattainment Areas

Pollutant	*Area classification*	*Primary standard attainment date*	*Major source threshold*	*Required offset ratio*
Ozone	Marginal	11/15/93	100 tpy VOCs or NO_x [a]	1.1:1
	Moderate	11/15/96	100 tpy VOCs or NO_x [a]	1.15:1
	Serious	11/15/99	50 tpy VOCs or NO_x [a]	1.2:1
	Severe-15	11/15/05	25 tpy VOCs or NO_x [a]	1.3:1[b]
	Severe-17	11/15/07	25 tpy VOCs or NO_x [a]	1.3:1[b]
	Extreme	11/15/10	10 tpy VOCs or NO_x [a]	1.5:1[b]
	Transport region not classified as serious, severe, or extreme	11/15/96	50 tpy VOCs 100 tpy NO_x	1.15:1
CO	Moderate	12/31/95	100 tpy CO	NA
	Serious	12/31/00	50 tpy CO[c]	NA
PM-10	Moderate	12/31/94[d]	100 tpy PM-10	NA
	Serious	12/31/01[e]	70 tpy PM-10	NA
SO_x	NA	11/15/95[f]	100 tpy SO_x	NA
NO_2	NA	11/15/95[f]	100 tpy NO_2	NA
Lead	NA	11/15/95[f]	100 tpy lead	NA

CO = carbon monoxide; NA = not applicable; NAAQS = national ambient air quality standards; NO_2 = nitrogen dioxide; NO_x = nitrogen oxides; PM-10 = particulate matter with an aerodynamic diameter less than or equal to 10 microns; SO_x = sulfur oxides; VOC = volatile organic compound.

[a]Under CAA Section 182(f), EPA may exempt certain major NO_x sources where NO_x reductions "would not contribute to attainment of the [NAAQS] in the area." Also, the "major source threshold" applies to VOC *or* NO_x emissions, but not to combined emissions.

[b]The required offset ratio is 1.2:1 for states that require all existing major sources in the nonattainment area to use BACT for control of VOC emissions.

[c]Stationary sources must contribute significantly to the ambient CO levels; that is, nonattainment cannot simply be the result of CO emissions from mobile sources.

[d]Or six years after area is redesignated to nonattainment, whichever is later.

[e]Or ten years after area is redesignated to nonattainment, whichever is later.

[f]Or five years after area is redesignated to nonattainment, whichever is later.

Source: Elsevier Science Inc.

source thresholds for nonattainment areas, which vary with nonattainment classification, are also listed in Table 3-4. These thresholds are used to determine applicability of the nonattainment NSR program, which regulates new and modified major sources in NAAQS nonattainment areas. Sources regulated under this program must obtain a nonattainment NSR preconstruction permit demonstrating that the source will meet the lowest achievable emission rate (LAER).

In addition, a source subject to nonattainment NSR must offset emission increases from its new facility with even larger emissions decreases—virtually no pollution is allowed from a new major source (or major modification) in a nonattainment area. As shown in Table 3-4, required offset ratios are substantially higher than 1:1 in the worst ozone nonattainment areas. For instance, sources in serious ozone nonattainment areas must decrease emissions elsewhere by 1.2 tons for every ton of increased emissions. In this way, net overall air quality is improved.

RACT

In their state implementation plans (SIPs), states must demonstrate how each nonattainment area will be brought into compliance. Many states attempt to achieve NAAQS compliance by requiring RACT. Usually, RACT is used to achieve the ozone NAAQS by imposing controls to reduce VOC and NO_x emissions. At a minimum, RACT must be imposed on all *existing* major stationary sources of VOC emissions located in ozone nonattainment areas that are classified as moderate or worse. (As noted earlier, new major sources are subject to the more stringent requirements of LAER.)

RACT is not clearly defined in the federal regulations. Consequently, requirements for a given source category may vary from state to state. However, EPA has specified that control techniques guidelines (CTGs) and alternative control techniques (ACTs) issued by the agency for various types of pollutant sources will qualify as RACT. (States may also determine RACT on a case-by-case basis for individual facilities.) CTGs (and to a lesser extent ACTs) contain a wealth of technical information regarding typical facility emissions, potential control technologies, and costs. The status and availability of CTGs and ACTs are provided in Table 3-5.

NSR

The generic term "NSR" or "new source review" actually refers to two separate CAA programs that both regulate cer-

TABLE 3-5
List of CTGs and ACTs

Source category	*Date*	*EPA and NTIS number*	*Document availability*[1]
Final CTGs (VOC control documents)			
Aerospace coatings	06/97	NA	TTNWeb NTIS
SOCMI distillation and reactor operations	08/15/93	EPA 450/4-91-031 PB94-120227	NTIS
Draft CTGs (VOC control documents)			
Batch processes	12/27/93	NA	—[2]
Industrial wastewater[3]	12/29/93	EPA-453/D-93-056	TTNWeb
Offset lithography[3]	11/08/93	PB95-201422	NTIS
Shipbuilding	06/95	NA	TTNWeb NTIS
Volatile organic liquid storage tanks	12/02/93	EPA-453/D-93-057 PB94-128519	NTIS
Wood furniture	08/95	EPA-453/D-95-002	TTNWeb NTIS
ACTs (NO_x control documents)			
Availability/extent of NO_x controls	NA	EPA-450/3-92-004 PB92-158260	NTIS
Cement plants	NA	EPA-453/R-94-004 PB94-183522	TTNWeb NTIS

TABLE 3-5
List of CTGs and ACTs—Continued

Source category	*Date*	*EPA and NTIS number*	*Document availability*[1]
ACTs (NO_x control documents)—Continued			
Gas turbines	NA	EPA-453/R-93-007 PB93-156586	TTNWeb NTIS
Glass manufacturing	NA	EPA-53/R-94-037	TTNWeb NTIS
Internal combustion engines	NA	EPA-453/R-93-032 PB94-104494	TTNWeb NTIS
Iron and steel manufacturing	NA	EPA-453/R-94-065	TTNWeb NTIS
Nitric/acid plants	NA	EPA-450/3-91-026 PB92-132828	TTNWeb NTIS
Nonutility boilers	NA	EPA-453/R-94-022 PB94-177177	TTNWeb NTIS
Process heaters	NA	EPA-453/R-93-034 PB94-120235	TTNWeb
Utility boilers	NA	EPA-453/R-94-023 PB94-184186	TTNWeb NTIS
ACTs (VOC control documents)			
Air oxidation processes in synthetic organic chemical manufacturing industry	12/84	EPA-450/3-84-015 PB-85-164-275	TTNWeb NTIS
Auto refinishing	05/94	EPA-453/R-94-031 PB94-191699	TTNWeb NTIS
Batch processes	02/94	EPA-453/R-93-017	NTIS
Bulk gasoline plants	06/78	EPA-450/2-78-035 PB-276-722	TTNWeb NTIS
Gasoline tank trucks and vapor collection systems	12/78	EPA-450/2-78-051 PB-290-568	TTNWeb NTIS
Graphic arts—Retrogravure and flexography	12/78	EPA-450/2-78-033 PB-292-490	TTNWeb NTIS
Equipment leaks from natural gas/gasoline processing plants	06/78	EPA-450/2-83-007 PB-84-161-520	TTNWeb NTIS
Factory surface coating of flat wood paneling	12/77	EPA-450/2-78-032 PB-292-490	TTNWeb NTIS
Fugitive emissions from synthetic organic chemical polymer and resin manufacturing equipment	03/84	EPA-450/3-83-006 PB-84-161-520	TTNWeb NTIS
Industrial cleaning solvents	NA	EPA-453/R-94-015 PB94-156791	TTNWeb NTIS
Industrial wastewater	04/94	NA	TTNWeb
Large petroleum dry cleaners	09/82	EPA-450/3-82-009 PB-83-124-875	TTNWeb NTIS
Leaks from petroleum refinery equipment	06/78	EPA-450/2-78-036 PB-286-158	TTNWeb NTIS
Manufacture of high-density polyethylene, polypropylene, and polystyrene resins	11/83	EPA-450/3-83-008 PB-84-134-600	TTNWeb NTIS
Manufacture of pneumatic rubber tires	12/78	EPA-450/2-78-030 PB-290-557	TTNWeb NTIS
Manufacture of synthesized pharmaceutical products	12/78	EPA-450/2-78-029 PB-290-580	TTNWeb NTIS

TABLE 3-5
List of CTGs and ACTs—Continued

Source category	*Date*	*EPA and NTIS number*	*Document availability*[1]
ACTs (VOC control documents)—Continued			
Offset lithography	06/94	PB95-201018	TTNWeb NTIS
Perchloroethylene dry cleaning systems	12/78	EPA-450/2-78-050 PB-290-613	TTNWeb NTIS
Pesticide application	1993	EPA-453/R-92-011 PB93-199131	NTIS
Petroleum liquid storage in external floating roof tanks	12/78	EPA-450/2-78-047 PB-290-579	TTNWeb NTIS
Plastic parts	02/94	EPA-453/R-94-017 PB94-158086	NTIS
Refinery vacuum-producing systems, wastewater separators, and process unit turnarounds	10/77	EPA-450/2-77-025 PB-275-662	TTNWeb NTIS
Shipbuilding	04/94	EPA-453/R-94-032 PB94-181694	TTNWeb NTIS
Solvent metal cleaning	12/83	EPA-450/2-77-022 PB-274-557	TTNWeb NTIS
Stage I vapor control systems	11/75	NA	TTNWeb
Stationary sources	05/78	EPA-450/2-78-022 PB-284-804	TTNWeb NTIS
Storage of petroleum liquids in fixed roof tanks	12/77	EPA-450/2-77-036 PB-276-749	TTNWeb NTIS
Surface coating operations	11/76	EPA-450/2-76-028 PB-260-386	TTNWeb NTIS
Surface coating of cans, coils, paper, fabrics, automobiles, and light-duty trucks	05/77	EPA-450/2-77-008 PB-272-445	TTNWeb NTIS
Surface coating of insulation on magnet wire	12/77	EPA-450/2-77-033 PB-278-258	TTNWeb NTIS
Surface coating of large appliances	12/77	EPA-450/2-77-034 PB-278-259	TTNWeb NTIS
Surface coating of metal furniture	12/77	EPA-450/2-77-033 PB-278-257	TTNWeb NTIS
Surface coating of miscellaneous metal parts and products	06/78	EPA-450/2-78-015 PB-286-157	TTNWeb NTIS
Tank truck gasoline loading terminals	12/77	EPA-450/2-77-026 PB-275-060	TTNWeb NTIS
Use of cutback asphalt	12/77	EPA-450/2-77-037 PB-278-185	TTNWeb NTIS
Volatile organic liquid storage tanks	02/94	EPA-453/R-94-001 PB-94-152741	NTIS

ACT = alternative control techniques; CTC = Control Technology Center; CTG = control techniques guideline; NA = not available; NO_x = nitrogen oxides; NTIS = National Technical Information Service; TTNWeb = Office of Air Quality Planning and Standards Technology Transfer Network web site; SOCMI = synthetic organic chemical manufacturing industry; VOC = volatile organic compound.

[1]Documents available on the TTNWeb CTC at http://www.epa.gov/oar/oaqps/ctc. To obtain CTGs and ACTs from NTIS, call (703) 487-4650.

[2]Paper copies of this document are available. See the TTNWeb CTC for details.

[3]EPA has no plans to finalize this CTG.

Source: Elsevier Science Inc.; adapted from information available on the TTNWeb CTC.

tain new or reconstructed major sources of criteria pollutant emissions. The PSD program applies to sources in NAAQS attainment areas and imposes best available control technology (BACT) on regulated facilities; nonattainment NSR applies LAER to facilities in NAAQS nonattainment areas.

Sources subject to these two programs, regardless of whether they are located in an attainment or nonattainment area, include the following:

- New major sources;
- Existing major sources that plan to implement a significant modification; and
- Existing nonmajor sources that plan to implement a modification that, by itself, qualifies as a major source.

Figure 3-1 (page 3.10) summarizes the applicability of NSR provisions to sources in attainment and nonattainment areas. The figure can be helpful when determining if either the PSD or nonattainment NSR provisions apply.

Key deadlines under the PSD and nonattainment NSR programs are listed in Table 3-6 (page 3.11).

Additional References

For more information on the NAAQS, PSD, and nonattainment NSR programs, consult the following issues of *The Air Pollution Consultant* (*APC*) and companion publications:

- Chapter 4 of *Complying With Clean Air Act Regulations: Issues and Techniques* (a 1995 publication of Elsevier Science Inc.), which provides a detailed summary of the NAAQS, PSD, and nonattainment NSR programs;
- Page 4.1 of the July/August 1996 issue of *APC*, which details important changes that EPA proposed for both the PSD and nonattainment NSR programs on July 23, 1996 (61 *FR* 38250);
- Page 4.1 of the November/December 1997 issue of *APC*, which reviews the agency's July 18, 1997 final revisions to the ozone and particulate matter NAAQS;
- Page 2.53 of the July/August 1998 issue of *APC*, which describes EPA's latest change to the VOC-exempt chemicals list—the addition of methyl acetate;
- Page 2.6 of the September/October 1998 issue of *APC*, which lists the areas of the country that EPA has exempted from the one-hour ozone NAAQS;
- Page 2.12 of the September/October 1998 issue of *APC*, which described EPA's proposal of regulatory language for stringent new NO_x reduction requirements in most of the eastern and central United States; and
- The "Perspectives" section in each issue of *APC* (page 5.1), which frequently addresses PSD and NSR. Authored by Gary McCutchen of RTP Environmental Associates Inc., this collection of articles provides unique insight into these complex regulatory topics.

FIGURE 3-1

Determining Whether Nonattainment NSR or PSD Applies[1]

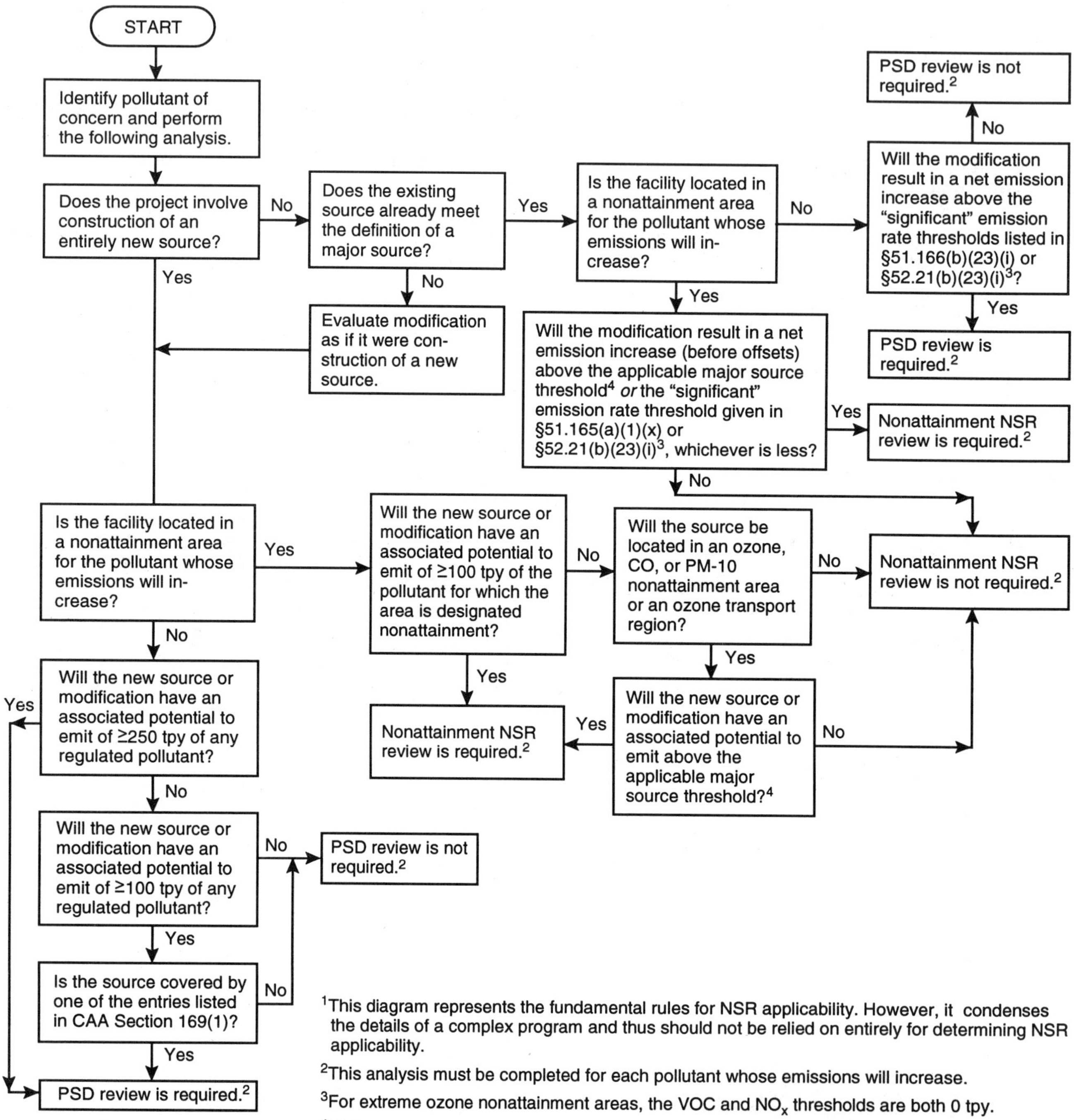

[1]This diagram represents the fundamental rules for NSR applicability. However, it condenses the details of a complex program and thus should not be relied on entirely for determining NSR applicability.

[2]This analysis must be completed for each pollutant whose emissions will increase.

[3]For extreme ozone nonattainment areas, the VOC and NO_x thresholds are both 0 tpy.

[4]Major source thresholds for ozone nonattainment areas vary and are listed in Table 3-4 on page 3.5. For serious CO nonattainment areas, the threshold is 50 tpy CO; for serious PM-10 nonattainment areas, the threshold is 70 tpy PM-10. For all other nonattainment areas, the threshold is 100 tpy.

Source: Elsevier Science Inc.

TABLE 3-6
Nonattainment NSR and PSD Regulatory Compliance Calendar[1]

Compliance date[2]		New source threshold date	Applicability and regulatory requirements[2]	CFR reference (Title 40)	FR reference	APC review[3]
Existing sources	New sources					
Nonattainment NSR Program						
NA	Upon start-up	NA	Sources located in NAAQS nonattainment areas that 1) construct or reconstruct a "major" source of criteria pollutants (or their precursors), or 2) complete a source modification that exceeds "significant-emission-rate values" must implement LAER for those emission sources *and* offset any increased emissions. An NSR permit is also required before source construction begins. Thresholds for major source determinations, significant-emission-rate values, and emission offset ratios depend on the severity of the area's nonattainment status (i.e., the area's classification).	Parts 51 and 52	—[4]	—[4]
PSD Program						
NA	Upon start-up	NA	Sources located in NAAQS attainment areas that 1) construct or reconstruct a "major" source of criteria pollutants (or their precursors), or 2) complete a source modification that exceeds "significant-emission-rate values" must implement BACT for those emission sources *and* ensure that increased emissions meet PSD guidelines for preserving ambient air quality (e.g., the increase does not consume more than the allowed PSD "increment"). A PSD permit is also required before source construction begins.	Parts 51 and 52	—[4]	—[4]

APC = *The Air Pollution Consultant*; BACT = best available control technology; LAER = lowest achievable emission rate; NA = not applicable; NAAQS = national ambient air quality standards; NSR = new source review; PSD = prevention of significant deterioration; RACT = reasonably available control technology.

[1]The calendar does *not* include rules and programs in the following four areas: 1) programs that are implemented on a state or local level where compliance deadlines are based on the individual agency's action (e.g., Title V operating permit programs and RACT requirements for sources in ozone nonattainment areas), 2) rules that impose deadlines on state or local government agencies (e.g., 40 *CFR* Part 58 regulations that require states and localities to establish ambient air quality monitoring systems), 3) rules that affect mobile sources, and 4) proposed rules that have not yet been promulgated.

[2]Various exemptions, extensions, and/or exceptions apply for some rules. These are *not* listed in this table.

[3]Refers to the volume, issue, and page number of the applicable *APC* review article. For example, 6:7, 4.1 refers to the *APC* article beginning on page 4.1 of Volume 6, Issue 7.

[4]Nonattainment NSR and PSD regulations were in place well before 1990. Thus, neither an *FR* reference nor a review article is available. Also, most of the changes to these programs imposed by the Clean Air Act Amendments of 1990 have been implemented through guidance documents. On July 23, 1996 (61 *FR* 38250), EPA proposed to codify many of these changes in 40 *CFR* Parts 51 and 52. A review of this proposal, which has not yet been finalized, may be found on page 4.1 of *APC* Volume 6, Issue 4.

Source: Elsevier Science Inc.

4 New Source Performance Standards

Another key program under Title I of the CAA is the new source performance standards (NSPS) program. NSPS regulations establish *technology-based* standards applicable to criteria pollutant emissions from certain stationary sources within specific categories. Some additional pollutants are also regulated including acid mist, fluorides, metals, and total reduced sulfur.

NSPS regulations apply to units irrespective of the area's attainment status. The NSPS program applies to new and modified stationary sources that meet the following criteria:

- The source includes one or more units (or groups of units) specified in an NSPS (Table 4-1 [page 4.2] identifies the industries, activities, and emission units for which NSPS regulations are currently in effect),
- Construction or reconstruction begins on or after a date specified in the NSPS for the applicable source category (usually the date on which the standard was proposed or promulgated), and
- Unit capacities meet or exceed those stated in the applicable NSPS.

NSPS regulations are codified in 40 *CFR* Part 60. Within this section of the *Code of Federal Regulations*, Subpart A, the so-called general NSPS provisions, deals with certain features that are common to source category-specific NSPS (e.g., performance testing, monitoring, etc.). Subpart A also contains references to states that have obtained authority to administer NSPS on EPA's behalf.

Subpart B of Part 60 deals with guideline documents that EPA must publish in conjunction with the proposal of a given NSPS. Subpart C references emission guidelines codified later in Part 60. (Emission guidelines are special rules that define minimum state programs for regulating pollutants from *existing* sources within the sulfuric acid production and waste combustion source categories.) The remaining subparts of Part 60 contain the source category-specific NSPS or emission guidelines. This currently includes Subparts Ca–VVV and regulates approximately 70 source categories.

A listing of the contents of Part 60, including the general provisions of Subparts A–C and the category-specific NSPS are presented in Table 4-2 (pages 4.3–4.5). In turn, important compliance deadlines applicable to NSPS-regulated facilities appear in Table 4-3 (pages 4.6–4.8).

Further information on the NSPS program can be found on page 4.5 of the May/June 1997 *The Air Pollution Consultant* and in Chapter 5 of the 1995 Elsevier Science Inc. publication *Complying With Clean Air Act Regulations: Issues and Techniques*. APC

©Elsevier Science Inc. 0-444-10032-6/98/$19.00+$5.50

TABLE 4-1

Industries, Activities, and Emission Units Subject to NSPS[1]

Aluminum reduction plants (primary)
Ammonium sulfate manufacturing
Asphalt:
—Hot mix facilities
—Processing and roofing manufacturing
Brass and bronze production plants (secondary)
Bulk gasoline terminals
Calciners and dryers in mineral industries
Coal preparation plants
Copper smelters (primary)
Ferroalloy production facilities
Flexible vinyl and urethane coating and printing
Gas turbines (stationary)
Glass manufacturing plants
Grain elevators
Incinerators
Kraft pulp mills
Lead-acid battery manufacturing plants
Lead smelters (primary and secondary)
Lime manufacturing plants
Magnetic tape coating facilities
Medical waste incinerators
Mineral processing plants (metallic and nonmetallic)
Municipal solid waste landfills
Municipal waste combustors
Natural gas processing plants
Nitric acid plants
Petroleum dry cleaners
Petroleum refineries, including wastewater systems and equipment leaks
Phosphate fertilizer industry:
—Diammonium phosphate plants
—Granular triple superphosphate storage facilities
—Superphosphoric acid plants
—Triple superphosphate plants
—Wet-process phosphoric acid plants
Phosphate rock plants
Polymer manufacturing industry
Polymeric coating of supporting substrates facilities
Portland cement plants
Publication rotogravure printing
Residential wood heaters (new)
Rubber tire manufacturing industry
Sewage treatment plants
Steam-generating activities:
—Electric utility units
—Fossil fuel-fired generators
—Industrial-commercial-institutional units
Steelmaking facilities:
—Argon oxygen decarburization vessels
—Electric arc furnaces
—Basic oxygen processes
Storage vessels:
—Petroleum liquids
—Volatile organic liquids
Sulfuric acid plants
Surface coating operations:
—Automobiles and light-duty trucks
—Beverage cans
—Large appliances
—Metal coils
—Metal furniture
—Plastic parts for business machines
—Pressure-sensitive tapes and labels
Synthetic fiber production facilities
SOCMI:
—Reactor processes
—Air oxidation unit processes
—Distillation operations
—Equipment leaks
Wool fiberglass insulation manufacturing plants
Zinc smelters (primary)

NSPS = new source performance standards; SOCMI = synthetic organic chemical manufacturing industry.

[1]Effective September 1998.

Source: 40 *CFR* Part 60.

TABLE 4-2
Key NSPS Regulations Found in 40 *CFR* Part 60

Part/Subpart	*Title*
Part 60	Standards of Performance for New Stationary Sources
Subpart A	• General Provisions
§60.1	—Applicability
§60.2	—Definitions
§60.3	—Units and abbreviations
§60.5	—Determination of construction or modification
§60.7	—Notification and recordkeeping
§60.8	—Performance tests
§60.11	—Compliance with standards and maintenance requirements
§60.13	—Monitoring requirements
§60.14	—Modification
§60.15	—Reconstruction
§60.18	—General control device requirements
§60.19	—General notification and reporting requirements
Subpart B	• Adoption and Submittal of State Plans for Designated Facilities
§60.20	—Applicability
§60.21	—Definitions
§60.22	—Publication of guideline documents, emission guidelines, and final compliance times
§60.24	—Emission standards and compliance schedules
C	• Emission Guidelines and Compliance Times
Ca	• [Reserved]
Cb	• Emission Guidelines and Compliance Times for Municipal Waste Combustors That Are Constructed on or Before December 19, 1995
Cc	• Emission Guidelines and Compliance Times for Municipal Solid Waste Landfills
Cd	• Emission Guidelines and Compliance Times for Sulfuric Acid Production Units
Ce	• Emission Guidelines and Compliance Times for Hospital/Medical/Infectious Waste Incinerators
D	• Standards of Performance for Fossil Fuel-Fired Steam Generators for Which Construction Is Commenced After August 17, 1971
Da	• Standards of Performance for Electric Utility Steam Generating Units for Which Construction Is Commenced After September 18, 1978
Db	• Standards of Performance for Industrial-Commercial-Institutional Steam Generating Units
Dc	• Standards of Performance for Small Industrial-Commercial-Institutional Steam Generating Units
E	• Standards of Performance for Incinerators
Ea	• Standards of Performance for Municipal Waste Combustors for Which Construction Is Commenced After December 20, 1989 and on or Before September 20, 1994
Eb	• Standards of Performance for Municipal Waste Combustors for Which Construction Is Commenced After September 20, 1994
Ec	• Standards of Performance for Hospital/Medical/Infectious Waste Incinerators for Which Construction Is Commenced After June 20, 1996
F	• Standards of Performance for Portland Cement Plants
G	• Standards of Performance for Nitric Acid Plants
H	• Standards of Performance for Sulfuric Acid Plants

TABLE 4-2
Key NSPS Regulations Found in 40 *CFR* Part 60—Continued

Part/Subpart	*Title*
I	• Standards of Performance for Hot Mix Asphalt Facilities
J	• Standards of Performance for Petroleum Refineries
K	• Standards of Performance for Storage Vessels for Petroleum Liquids for Which Construction, Reconstruction, or Modification Commenced After June 11, 1973, and Prior to May 19, 1978
Ka	• Standards of Performance for Storage Vessels for Petroleum Liquids for Which Construction, Reconstruction, or Modification Commenced After May 18, 1978, and Prior to July 23, 1984
Kb	• Standards of Performance for Volatile Organic Liquid Storage Vessels (Including Petroleum Liquid Storage Vessels) for Which Construction, Reconstruction, or Modification Commenced after July 23, 1984
L	• Standards of Performance for Secondary Lead Smelters
M	• Standards of Performance for Secondary Brass and Bronze Production Plants
N	• Standards of Performance for Primary Emissions From Basic Oxygen Process Furnaces for Which Construction Is Commenced After June 11, 1973
Na	• Standards of Performance for Secondary Emissions From Basic Oxygen Process Steelmaking Facilities for Which Construction Is Commenced After January 20, 1983
O	• Standards of Performance for Sewage Treatment Plants
P	• Standards of Performance for Primary Copper Smelters
Q	• Standards of Performance for Primary Zinc Smelters
R	• Standards of Performance for Primary Lead Smelters
S	• Standards of Performance for Primary Aluminum Reduction Plants
T	• Standards of Performance for the Phosphate Fertilizer Industry: Wet-Process Phosphoric Acid Plants
U	• Standards of Performance for the Phosphate Fertilizer Industry: Superphosphoric Acid Plants
V	• Standards of Performance for the Phosphate Fertilizer Industry: Diammonium Phosphate Plants
W	• Standards of Performance for the Phosphate Fertilizer Industry: Triple Superphosphate Plants
X	• Standards of Performance for the Phosphate Fertilizer Industry: Granular Triple Superphosphate Storage Facilities
Y	• Standards of Performance for Coal Preparation Plants
Z	• Standards of Performance for Ferroalloy Production Facilities
AA	• Standards of Performance for Steel Plants: Electric Arc Furnaces Constructed After October 21, 1974 and on or Before August 17, 1983
AAa	• Standards of Performance for Steel Plants: Electric Arc Furnaces and Argon-Oxygen Decarburization Vessels Constructed After August 7, 1983
BB	• Standards of Performance for Kraft Pulp Mills
CC	• Standards of Performance for Glass Manufacturing Plants
DD	• Standards of Performance of Grain Elevators
EE	• Standards of Performance for Surface Coating of Metal Furniture
FF	• [Reserved]
GG	• Standards of Performance for Stationary Gas Turbines
HH	• Standards of Performance for Lime Manufacturing Plants
KK	• Standards of Performance for Lead-Acid Battery Manufacturing Plants
LL	• Standards of Performance for Metallic Mineral Processing Plants

©Elsevier Science Inc. 0-444-10032-6/98/$19.00+$5.50

TABLE 4-2
Key NSPS Regulations Found in 40 *CFR* Part 60—Continued

Part/Subpart	*Title*
MM	• Standards of Performance for Automobile and Light-Duty Truck Surface Coating Operations
NN	• Standards of Performance for Phosphate Rock Plants
PP	• Standards of Performance for Ammonium Sulfate Manufacture
QQ	• Standards of Performance for the Graphic Arts Industry: Publication Rotogravure Printing
RR	• Standards of Performance for Pressure Sensitive Tape and Label Surface Coating Operations
SS	• Standards of Performance for Industrial Surface Coating: Large Appliances
TT	• Standards of Performance for Metal Coil Surface Coating
UU	• Standards of Performance for Asphalt Processing and Asphalt Roofing Manufacture
VV	• Standards of Performance for Equipment Leaks of VOC in the SOCMI
WW	• Standards of Performance for the Beverage Can Surface Coating Industry
XX	• Standards of Performance for Bulk Gasoline Terminals
AAA	• Standards of Performance for New Residential Wood Heaters
BBB	• Standards of Performance for the Rubber Tire Manufacturing Industry
CCC	• [Reserved]
DDD	• Standards of Performance for VOC Emissions From the Polymer Manufacturing Industry
EEE	• [Reserved]
FFF	• Standards of Performance for Flexible Vinyl and Urethane Coating and Printing
GGG	• Standards of Performance for Equipment Leaks of VOC in Petroleum Refineries
HHH	• Standards of Performance for Synthetic Fiber Production Facilities
III	• Standards of Performance for VOC Emissions From the SOCMI Air Oxidation Unit Processes
JJJ	• Standards of Performance for Petroleum Dry Cleaners
KKK	• Standards of Performance for Equipment Leaks of VOC From Onshore Natural Gas Processing Plants
LLL	• Standards of Performance for Onshore Natural Gas Processing; SO_2 Emissions
MMM	• [Reserved]
NNN	• Standards of Performance for VOC Emissions From Synthetic Organic Chemical Manufacturing Industry Distillation Operations
OOO	• Standards of Performance for Nonmetallic Mineral Processing Plants
PPP	• Standard of Performance for Wool Fiberglass Insulation Manufacturing Plants
QQQ	• Standards of Performance for VOC Emissions From Petroleum Refinery Wastewater Systems
RRR	• Standards of Performance for VOC Emissions From SOCMI Reactor Processes
SSS	• Standards of Performance for Magnetic Tape Coating Facilities
TTT	• Standards of Performance for Industrial Surface Coating: Surface Coating of Plastic Parts for Business Machines
UUU	• Standards of Performance for Calciners and Dryers in Mineral Industries
VVV	• Standards of Performance for Polymeric Coating of Supporting Substrates Facilities
WWW	• Standards of Performance for Municipal Solid Waste Landfills

NSPS = new source performance standards; SO_2 = sulfur dioxide; SOCMI = synthetic organic chemical manufacturing industry; VOC = volatile organic compound.

Source: Elsevier Science Inc.

TABLE 4-3
NSPS Regulatory Compliance Calendar[1]

Compliance date[2]		New source threshold date[3]	Applicability and regulatory requirements[2]	CFR reference (Title 40)	FR reference	APC review[4]
Existing sources	New sources					
Automobile and light-duty truck surface coating NSPS						
NA	Within 180 days of start-up or 60 days of achieving maximum production rate, whichever is earlier	10/05/79	Automobile and light-duty truck manufacturers must limit the emission of VOCs from their surface coating operations; emission limits vary depending on the facility's turnover rate (a measure of production) and the volume of coating solids applied.	Part 60 Subpart MM	10/11/94 59 *FR* 51383	5:1, 2.50
Industrial, commercial, and institutional steam boilers						
NA	Within 180 days of start-up or 60 days of achieving maximum production rate, whichever is earlier	07/09/97	Fossil fuel-fired industrial, commercial, and institutional utility boilers with input heat capacity ≥29 MW (100 MMBtu/hr) must limit their emissions of NO_x, particulate matter, and SO_2	Part 60 Subpart Db		—[5]
Mineral industry calciner and dryer NSPS						
NA	Within 180 days of start-up or 60 days of achieving maximum production rate, whichever is earlier	04/23/86	Plants that process any of the minerals listed in the definition of mineral processing plant (§60.731) must implement particulate controls for their calciners and dryers.	Part 60, Subpart UUU	9/28/92 57 *FR* 44496 7/29/93 58 *FR* 40591	3:1, 2.13 (original rule) 3:6, 2.30 (corrections)
SOCMI reactor process NSPS						
NA	Within 180 days of start-up or 60 days of achieving maximum production rate, whichever is earlier	06/29/90	SOCMI reactor processes that produce one of the chemicals listed in §60.707 must implement BDT for those processes.	Part 60, Subpart RRR	8/31/93 58 *FR* 45948 11/27/95 60 *FR* 58237	3:6, 2.7 (original rule) 6:2, 2.53 (corrections)

TABLE 4-3
NSPS Regulatory Compliance Calendar[1]—Continued

Compliance date[2]: Existing sources	Compliance date[2]: New sources	New source threshold date[3]	Applicability and regulatory requirements[2]	CFR reference (Title 40)	FR reference	APC review[4]
Utility steam boilers						
NA	Within 180 days of start-up or 60 days of achieving maximum production rate, whichever is earlier	07/09/97	Fossil fuel-fired utility boilers with input heat capacity ≥73 MW (250 MMBtu/hr) must limit their emissions of NO_x, particulate matter, and SO_2	Part 60 Subpart Da		—[5]
NSPS and emission guidelines for MSW landfills						
—[6]	Start-up or 3/12/99, whichever is later	05/30/91	*New* MSW landfills with capacity ≥2.75 million tons that emit >55 tpy NMOCs must implement MACT-based controls; *existing* MSW landfills with capacity ≥2.75 million tons that emit >55 tpy NMOCs must implement state program (MACT-based) controls.	Part 60, Subparts Cc and WWW	3/12/96 61 *FR* 9905	6:4, 2.30
NSPS and emission guidelines for MWCs						
—[6]	Within 180 days of start-up or 60 days of achieving maximum production rate, whichever is earlier	09/20/94[7]	*New* MWCs with waste burning capacity >250 tons/day must implement MACT-based controls; *existing* MWCs with waste burning capacity >250 tons/day must implement state program (MACT-based) controls.	Part 60, Subparts Cb and Eb	12/19/95 60 *FR* 65387 8/25/97 62 *FR* 45124	6:2, 2.12 (original rule) 7:6, 2.27 (applicability revision)

TABLE 4-3
NSPS Regulatory Compliance Calendar[1]—Continued

Compliance date[2] *Existing sources*	*Compliance date*[2] *New sources*	*New source threshold date*[3]	*Applicability and regulatory requirements*[2]	*CFR reference (Title 40)*	*FR reference*	*APC review*[4]
NSPS and emission guidelines for medical/infectious/hospital waste incinerators						
—[6]	Within 180 days of start-up or 60 days of achieving maximum production rate, whichever is earlier	06/20/96[7]	All *new* nonexempt incinerators that combust medical/infectious/hospital waste must implement MACT-based controls; *existing* medical waste incinerators must implement state program (MACT-based) controls.	Part 60, Subparts Ce and Ec	9/15/97 62 *FR* 48348	8:1,2.1

APC = *The Air Pollution Consultant*; BDT = best demonstrated technology; MACT = maximum achievable control technology; MSW = municipal solid waste; MWC = municipal waste combustor; NA = not applicable; NMOC = nonmethane organic compound; NO_x = nitrogen oxides; NSPS = new source performance standards; RACT = reasonably available control technology; SO_2 = sulfur dioxide; SOCMI = synthetic organic chemical manufacturing industry; VOC = volatile organic compound.

[1]The calendar does *not* include rules and programs in the following five areas: 1) programs that are implemented on a state or local level where compliance deadlines are based on the individual agency's action (e.g., Title V operating permit programs and RACT requirements for sources in ozone nonattainment areas); 2) rules that impose deadlines on state or local government agencies (e.g., 40 *CFR* Part 58 regulations that require states and localities to establish ambient air quality monitoring systems); 3) rules promulgated prior to May 8, 1991; 4) rules that affect mobile sources; and 5) proposed rules that have not yet been promulgated.

[2]Various exemptions, extensions, and/or exceptions apply for some rules. These are *not* listed in this table.

[3]New sources are those for which construction or reconstruction began after the listed threshold date.

[4]Refers to the volume, issue, and page number of the applicable *APC* review article. For example, 5:1, 2.50 refers to the *APC* article beginning on page 2.50 of Volume 5, Issue 1.

[5]A review article of the September 16, 1998 rule modifying this NSPS will appear in the January/February 1999 issue of *APC*.

[6]Existing source compliance deadlines are dependent on state programs implementing the Part 60 emission guidelines; thus, they are not given in this table.

[7]Or March 16, 1998 for *modified* medical/infectious/hospital waste incinerators and June 19, 1996 for *modified* MWCs.

Source: Elsevier Science Inc.

5 Hazardous Air Pollutant Control Programs

In enacting Title III of the Clean Air Act Amendments of 1990, Congress initiated a two-pronged attack on emissions of hazardous air pollutants (HAPs). First, under the existing NESHAP program, the legislature directed EPA to issue technology-based control standards to control routine HAP emissions—these are called MACT standards. Second, the agency was directed to promulgate new control regulations to address the accidental release of HAPs—this is the accidental release prevention program.

To facilitate access to the Title III information presented in this guide, this chapter is divided into three sections: 1) NESHAP and MACT applicability and general information, which begins just below; 2) the general NESHAP provisions, which begin on page 5.26; and 3) the accidental release prevention program, which also begins on page 5.26.

NESHAP and MACT Applicability and General Information

When Congress amended the CAA in 1970, EPA was directed to determine which HAPs to regulate and then set appropriate standards to provide an "ample margin of safety to protect human health." This health-based approach turned out to be a slow process because the statutory criterion was difficult to define and even harder to implement. Over a 20-year period, EPA was able to promulgate fewer than two dozen control standards regulating just eight toxic air pollutants. These pre-1990 amendment rules, which are often referred to simply as "NESHAPs" (in contrast to post-1990 amendment HAP control rules, which are called MACT standards), are codified in 40 *CFR* Part 61. A listing of the NESHAPs, which EPA continues to administer and enforce, can be found in Table 5-1 (page 5.2).

Due to the ineffectiveness of the original NESHAP program, a more aggressive approach was taken in the 1990 amendments. Rather than wait for EPA to evaluate individual HAPs and establish health-based standards for them, Congress specified 189 compounds that must be regulated and laid out a specific regulatory framework for EPA to follow. The list of HAPs is presented in Table 5-2 (pages 5.3–5.5). (This list actually contains only 188 chemicals; based on an industry petition, caprolactam was removed from the HAP list by EPA on June 18, 1996 [61 *FR* 30816].) HAPs ranked by annual emissions and their top 20 sources are presented in Tables 5-3 and 5-4, respectively (page 5.6).

The new control scheme for HAPs included in Title III of the 1990 amendments mandates that EPA set technology-based standards for source categories that emit HAPs. Because these standards are designed to require the best demonstrated control for a given type of facility, they are known as MACT standards. The status of MACT standard promulgation is outlined in Table 5-5 (pages 5.7–5.12), which includes all source categories for which MACT standards are to be issued and corresponding promulgation deadlines. For MACT standards that have been proposed and/or finalized, the issuance date, the applicable 40 *CFR* Part 63 subpart, and a reference to the review article in *APC* are also provided.

New and existing "major sources" that fall within listed source categories (i.e., those listed in Table 5-5) are subject to the MACT standards. In addition, some nonmajor (i.e., area) sources may also be subject to MACT or other control requirements. Under Title III, a major source is defined as "any stationary source or group of stationary sources located within a contiguous area and under common control that emits or has the potential to emit considering controls, in the aggregate, 10 tons per year or more of any [HAP] or 25 tons per year or more of any combination of [HAPs] [CAA Section 112(a)]." (This major source threshold is different from new source review and other CAA program thresholds.)

As a result of the new MACT standard requirements, regulated sources are subject to increased monitoring and stack testing requirements. A list of recommended HAP emission sampling and analytical methods is presented in Table 5-6 (pages 5.13–5.16).

Finally, Table 5-7 (pages 5.17–5.25) presents a regulatory compliance calendar for sources regulated under the NESHAP program. Note that this table is particularly important, in as much as the existing source compliance deadlines for a number of important MACT standards fall in 1999.

Besides issuing MACT standards, EPA has also been directed by Congress to issue various related rulemakings and report the results of prescribed research to Congress. Information regarding these requirements can be found in the following *APC* articles:

- Page 2.34 of the March/April 1993 *APC*, which details how sources may voluntarily reduce their HAP emissions by 90% (95% for particulate HAP) and thereby earn a 6-year compliance extension from applicable MACT standards; this is called HAP early reduction and was issued by EPA on December 29, 1992 (57 *FR* 61970).
- Page 2.11 of the September/October 1994 *APC*, which reviews a May 20, 1994 rule (59 *FR* 26429) wherein EPA describes how it will require states to prepare case-by-case MACT if the federal agency fails to meet one of the so-called "MACT hammers" (i.e., promulgation dead-

©Elsevier Science Inc. 0-444-10032-6/98/$19.00+$5.50

lines for issuing MACT standards). Note that EPA has avoided MACT hammers scheduled in 1994 and 1996, but will be hard pressed to avoid the next one, which falls on May 15, 1999.

- Page 4.6 of the March/April 1996 *APC*, which provides details on some of the other miscellaneous requirements of CAA Title III.
- Page 2.2 of the March/April 1997 *APC*, which reports on the Section 112(g) MACT preconstruction review program promulgated by EPA on December 27, 1996 (61 *FR* 68384).
- Page 2.25 of the July/August 1998 issue of *APC*, which describes the agency's plan to add aviation fuel distribution and open scrap tire burning to the list of source categories for which MACT standards will be issued; this is based on the mandate of CAA Section 112(c)(6), which directs EPA to examine the emissions of seven particular classes of HAPs.
- Page 2.54 of the July/August 1998 issue of *APC*, announcing EPA's finding in terms of the adequacy of its existing authority to address HAP deposition to the "great waters" of the United States under the mandate of CAA Section 112(m).

TABLE 5-1
Original NESHAP Regulations Issued Under Part 61

Subpart	*Regulatory provision*
A	**General provisions**
	National emission standards for:
B	• Radon emissions from underground uranium mines
C	• Beryllium
D	• Beryllium rocket motor firing
E	• Mercury
F	• Vinyl chloride
H	• Radionuclides other than radon from Department of Energy facilities
I	• Radionuclide emissions from facilities licensed by the Nuclear Regulatory Commission and federal facilities not covered by Subpart H
J	• Equipment leaks (fugitive emission sources) of benzene
K	• Radionuclide emissions from elemental phosphorus plants
L	• Benzene emissions from coke by-product recovery plants
M	• Asbestos
N	• Inorganic arsenic emissions from glass manufacturing plants
O	• Inorganic arsenic emissions from primary copper smelters
P	• Inorganic arsenic emissions from arsenic trioxide and metallic arsenic production facilities
Q	• Radon emissions from Department of Energy facilities
R	• Radon emissions from phosphogypsum stacks
T	• Radon emissions from the disposal of uranium mill tailings
V	• Equipment leaks (fugitive emission sources)
W	• Radon emissions from operating mill tailings
Y	• Benzene emissions from benzene storage vessels
BB	• Benzene emissions from benzene transfer operations
FF	• Benzene waste operations

NESHAP = national emission standards for hazardous air pollutants.

Source: Elsevier Science Inc.

TABLE 5-2
HAPs Regulated Under the NESHAP Program[1]

Chemical name	*CAS no.*
Acetaldehyde	75-07-0
Acetamide	60-35-5
Acetonitrile	75-05-8
Acetophenone	98-86-2
2-Acetylaminofluorene	53-96-3
Acrolein	107-02-8
Acrylamide	79-06-1
Acrylic acid	79-10-7
Acrylonitrile	107-13-1
Allyl chloride	107-05-1
4-Aminobiphenyl	92-67-1
Aniline	62-53-3
o-Anisidine	90-04-0
Asbestos	1332-21-4
Benzene (including benzene from gasoline)	71-43-2
Benzidine	92-87-5
Benzotrichloride	98-07-7
Benzyl chloride	100-44-7
Biphenyl	92-52-4
Bis(chloromethyl)ether	542-88-1
Bis(2-ethylhexyl)phthalate (DEHP)	117-81-7
Bromoform	75-25-2
1,3-Butadiene	106-99-0
Calcium cyanamide	156-62-7
Captan	133-06-2
Carbaryl	63-25-2
Carbon disulfide	75-15-0
Carbon tetrachloride	56-23-5
Carbonyl sulfide	463-58-1
Catechol	120-80-9
Chloramben	133-90-4
Chlordane	57-74-9
Chlorine	7782-50-5
Chloroacetic acid	79-11-8
2-Chloroacetophenone	532-27-4
Chlorobenzene	108-90-7
Chlorobenzilate	510-15-6
Chloroform	67-66-3
Chloromethyl methyl ether	107-30-2
Chloroprene	126-99-8
Cresols/cresylic acid (isomers and mixture)	1319-77-3
m-Cresol	108-39-4
o-Cresol	95-48-7
p-Cresol	106-44-5
Cumene	98-82-8
2,4-D (salts and esters)	94-75-7
DDE	3547-04-4
Diazomethane	334-88-3
Dibenzofurans	132-64-9
1,2-Dibromo-3-chloropropane	96-12-8
Dibutylphthalate	84-74-2
1,4-Dichlorobenzene(p)	106-46-7
3,3-Dichlorobenzidene	91-94-1
Dichloroethyl ether [Bis(2-chloroethyl)ether]	111-44-4
1,3-Dichloropropene	542-75-6
Dichlorvos	62-73-7
Diethanolamine	111-42-2
N,N-Diethyl aniline (N,N-Dimethyl-aniline)	121-69-7
Diethyl sulfate	64-67-5
3,3-Dimethoxybenzidine	119-90-4
Dimethyl aminoazobenzene	60-11-7
3,3′-Dimethyl benzidine	119-93-7
Dimethyl carbamoyl chloride	79-44-7
Dimethyl formamide	68-12-2
1,1-Dimethyl hydrazine	57-14-7
Dimethyl phthalate	131-11-3
Dimethyl sulfate	77-78-1
4,6-Dinitro-*o*-cresol	534-52-1
2,4-Dinitrophenol	51-28-5
2,4-Dinitrotoluene	121-14-2
1,4-Dioxane (1,4-Diethyleneoxide)	123-91-1
1,2-Diphenylhydrazine	122-66-7

TABLE 5-2
HAPs Regulated Under the NESHAP Program[1]—Continued

Chemical name	CAS no.
Epichlorohydrin (l-Chloro-2,3-epoxypropane)	106-89-8
1,2-Epoxybutane	106-88-7
Ethyl acrylate	140-88-5
Ethyl benzene	100-41-4
Ethyl carbamate (Urethane)	51-79-6
Ethyl chloride (Chloroethane)	75-00-3
Ethylene dibromide (Dibromoethane)	106-93-4
Ethylene dichloride (1,2-Dichloro-ethane)	107-06-2
Ethylene glycol	107-21-1
Ethylene imine (Aziridine)	151-56-4
Ethylene oxide	75-21-8
Ethylene thiourea	96-45-7
Ethylidene dichloride (1,1-Dichloroethane)	75-34-3
Formaldehyde	50-00-0
Heptachlor	76-44-8
Hexachlorobenzene	118-74-1
Hexachlorobutadiene	87-68-3
Hexachlorocyclopentadiene	77-47-4
Hexachloroethane	67-72-1
Hexamethylene-1,6-diisocyanate	822-06-0
Hexamethylphosphoramide	680-31-9
Hexane	110-54-3
Hydrazine	302-01-2
Hydrochloric acid	7647-01-0
Hydrogen fluoride (Hydrofluoric acid)	7664-39-3
Hydroquinone	123-31-9
Isophorone	78-59-1
Lindane (all isomers)	58-89-9
Maleic anhydride	108-31-6
Methanol	67-56-1
Methoxychlor	72-43-5
Methyl bromide (Bromomethane)	74-83-9
Methyl *tert*-butyl ether	1634-04-4
Methyl chloride (Chloromethane)	74-87-3
Methyl chloroform (1,1,1-Trichloroethane)	71-55-6
Methyl ethyl ketone (2-Butanone)	78-93-3
Methyl hydrazine	60-34-4
Methyl iodide (Iodomethane)	74-88-4
Methyl isobutyl ketone (Hexone)	108-10-1
Methyl isocyanate	624-83-9
Methyl methacrylate	80-62-6
4,4-Methylene bis(2-chloroaniline)	101-14-4
Methylene chloride (Dichloromethane)	75-09-2
Methylene diphenyl diisocyanate (MDI)	101-68-8
4,4′-Methylenedianiline	101-77-9
Naphthalene	91-20-3
Nitrobenzene	98-95-3
4-Nitrobiphenyl	92-93-3
4-Nitrophenol	100-02-7
2-Nitropropane	79-46-9
N-Nitroso-N-methylurea	684-93-5
N-Nitrosodimethylamine	62-75-9
N-Nitrosomorpholine	59-89-2
Parathion	56-38-2
Pentachloronitrobenzene (Quinto-benzene)	82-68-8
Pentachlorophenol	87-86-5
Phenol	108-95-2
p-Phenylenediamine	106-50-3
Phosgene	75-44-5
Phosphine	7803-51-2
Phosphorus	7723-14-0
Phthalic anhydride	85-44-9
Polychlorinated biphenyls (Aroclors)	1336-36-3
1,3-Propane sultone	1120-71-4
beta-Propiolactone	57-57-8
Propionaldehyde	123-38-6
Propoxur (Baygon)	114-26-1
Propylene dichloride (1,2-Dichloropropane)	78-87-5
Propylene oxide	75-56-9
1,2-Propylenimine (2-Methyl azirid-ine)	75-55-8

TABLE 5-2
HAPs Regulated Under the NESHAP Program[1]—Continued

Chemical name	*CAS no.*
Quinoline	91-22-5
Quinone	106-51-4
Styrene	100-42-5
Styrene oxide	96-09-3
2,3,7,8-Tetrachlorodibenzo-*p*-dioxin	1746-01-6
1,1,2,2-Tetrachloroethane	79-34-5
Tetrachloroethylene (Perchloroethylene)	127-18-4
Titanium tetrachloride	7550-45-0
Toluene	108-88-3
2,4-Toluene diamine	95-80-7
2,4-Toluene diisocyanate	584-84-9
o-Toluidine	95-53-4
Toxaphene (Chlorinated camphene)	8001-35-2
1,2,4-Trichlorobenzene	120-82-1
1,1,2-Trichloroethane	79-00-5
Trichloroethylene	79-01-6
2,4,5-Trichlorophenol	95-95-4
2,4,6-Trichlorophenol	88-06-2
Triethylamine	121-44-8
Trifluralin	1582-09-8
2,2,4-Trimethylpentane	540-84-1
Vinyl acetate	108-05-4
Vinyl bromide	593-60-2
Vinyl chloride	75-01-4
Vinylidene chloride (1,1-Dichloroethylene)	75-35-4
Xylenes (isomers and mixtures)	1330-20-7
m-Xylenes	108-38-3
o-Xylenes	95-47-6
p-Xylenes	106-42-3
Antimony compounds[2]	—
Arsenic compounds (inorganic including arsine)[2]	—
Beryllium compounds[2]	—
Cadmium compounds[2]	—
Chromium compounds[2]	—
Cobalt compounds[2]	—
Coke oven emissions	—
Cyanide compounds[2,3]	—
Glycol ethers[2,4]	—
Lead compounds[2]	—
Manganese compounds[2]	—
Mercury compounds[2]	—
Fine mineral fibers[5]	—
Nickel compounds[2]	—
Polycylic organic matter[6]	—
Radionuclides (including radon)[7]	—
Selenium compounds[2]	—

CAS = Chemical Abstract Service; HAP = hazardous air pollutant; NESHAP = national emission standards for hazardous air pollutants.

[1]The HAP list in CAA Section 112(b)(1) was modified on June 18, 1996 (61 *FR* 30816) by the removal of caprolactam.

[2]Unless otherwise specified, these listings include any unique chemical substance that contains the named chemical (i.e., antimony, arsenic, etc.) as part of that chemical's infrastructure.

[3]X′CN where X = H′ or any other group where a formal dissociation may occur, for example, KCN or $Ca(CN)_2$.

[4]Includes mono- and di-ethers of ethylene glycol, diethylene glycol, and triethylene glycol $R–(OCH_2CH_2)_n–OR'$ where:
n = 1, 2, or 3;
R = alkyl or aryl groups;
R′ = R, H, or groups that, when removed, yield glycol ethers with the structure $R–(OCH_2CH)_n–OH$; polymers are excluded from the glycol category.

[5]Includes mineral fiber emissions from facilities manufacturing or processing glass, rock, or slag fibers (or other mineral-derived fibers) with average diameter of 1 micrometer or less.

[6]Includes organic compounds that have more than one benzene ring and a boiling point greater than or equal to 100°C.

[7]A type of atom that spontaneously undergoes radioactive decay.

Source: CAA Section 112(b)(1) and §63.60.

TABLE 5-3
36 HAPs Ranked by Annual Emissions (1990 Data)[1]

Pollutant	Emissions (tons/yr)	Pollutant	Emissions (tons/yr)
Toluene	1,220,000	Ethylene dichloride	1,270
Polycyclic organic matter (POM)	753,000	Bis(2-chloroethyl) ether	790
Benzene	572,000	Cadmium compounds	790
Formaldehyde	281,000	Acrylonitrile	698
Xylenes	187,000	Ethylene oxide	652
1,3-Butadiene	123,000	Vinyl chloride	517
Tetrachloroethylene	109,000	Chromium compounds	294
Acetaldehyde	95,300	Methylene diphenyl diisocyanate (MDI)	273
Trichloroethylene	53,300	Mercury compounds	246
Acrolein	49,300	2,4-Toluene	45
Methylene chloride	43,400	Antimony compounds	21.8
Hydrazine	39,400	Ethylene dibromide	16.8
Glycol ethers	23,000	Acrylamide	14.4
Styrene	16,700	Beryllium compounds	9.29
Arsenic compounds	13,600	Phosgene	2.85
Chloroform	6,930	2,3,7,8-TCDF	0.0144
Nickel compounds	5,360	2,3,7,8-TCDD	0.00156
Lead compounds	3,760		
Manganese	1,700		

HAP = hazardous air pollutant; TCDD = tetrachlorodibenzo-*p*-dioxin; TCDF = tetrachlorodibenzofuran.

[1]These 36 HAPs represent about 86% of the total annual emissions of all 188 listed HAPs.

Source: Elsevier Science Inc.; adapted from EPA-454/R-96-007.

TABLE 5-4
Top 20 Sources of HAP Emissions (1993 Data)

Source category	Emissions (tons/yr)	Source category	Emissions (tons/yr)
On-road motor vehicles	1,389,000	Primary nonferrous metals production	38,000
Consumer and commercial product solvent use	414,000	Cellulose manmade fiber	38,000
Forest fires and wildfires	208,000	Petroleum refining	27,000
Glycol dehydrators	206,000	Municipal waste combustion	25,000
Non-road mobile vehicles	146,000	Motor vehicle surface coating	23,000
Prescribed burning	134,000	Gasoline distribution, Stage II	22,000
Residential wood combustion	99,000	Utility coal combustion	21,000
Dry cleaning	96,000	Plastic materials and resin manufacturing	21,000
Organic chemical manufacturing	91,000	Flexible polyurethane form production	20,000
Pulp and paper production	89,000		
Halogenated solvent cleaning	61,000		

HAP = hazardous air pollutant.

Source: Elsevier Science Inc.; adapted from EPA-454/R-97-013.

TABLE 5-5
Status of MACT Standards

Industry group and source category	*Deadline*	*Date proposed*	*Date promulgated*	*Part 63 subpart*	*APC review*[1,2]
Agricultural chemical production[3]					
Butadiene-furfural cotrimer (R-11) production	11/15/00	11/10/97		MMM	8:2, 2.6
		12/31/92	04/22/94	H, I[4]	4:3, 4.1
Captafol production	11/15/97	11/10/97		MMM	8:2, 2.6
		12/31/92	04/22/94	H, I[4]	4.3, 4.1
Captan production	11/15/97	11/10/97		MMM	8:2, 2.6
		12/31/92	04/22/94	H, I[4]	4.3, 4.1
4-Chloro-2-methylphenoxyacetic acid production	11/15/97	11/10/97		MMM	8:2, 2.6
Chloroneb production	11/15/97	11/10/97		MMM	8:2, 2.6
Chlorothalonil production	11/15/97	11/10/97		MMM	8:2, 2.6
		12/31/92	04/22/94	H, I[4]	4.3, 4.1
2,4-D salt and ester production	11/15/97	11/10/97		MMM	8:2, 2.6
Dacthal™ production	11/15/97	11/10/97		MMM	8:2, 2.6
		12/31/92	04/22/94	H, I[4]	4.3, 4.1
4,6-Dinitro-*o*-cresol production	11/15/97	11/10/97		MMM	8:2, 2.6
Sodium pentachlorophenate production	11/15/97	11/10/97		MMM	8:2, 2.6
Tordon™ acid production	11/15/97	11/10/97		MMM	8:2, 2.6
		12/31/92	04/22/94	H, I[4]	4.3, 4.1
Ferrous metal processing					
Coke by-product plants	11/15/00				
Coke ovens—charging, top side, and door leaks	12/31/92	04/23/87[5]	10/27/93	L	4:1, 2.29
Coke ovens—pushing, quenching, and battery stacks	11/15/00				
Ferroalloy production	11/15/97	08/04/98		XXX	8:6, 2.30
Integrated iron and steel manufacturing	11/15/00				
Iron foundries	11/15/00				
Steel foundries	11/15/00				
Steel pickling—HCl process	11/15/97	09/18/97		CCC	8:1, 2.27
Fiber production processes					
Acrylic fiber/modacrylic fiber production	11/15/97				
Rayon production	11/15/00				
Spandex production	11/15/00				
Food and agriculture processes					
Baker's yeast manufacturing	11/15/00				
Cellulose food casing manufacturing	11/15/00				
Vegetable oil production	11/15/00				

TABLE 5-5
Status of MACT Standards—Continued

Industry group and source category	*Deadline*	*Date proposed*	*Date promulgated*	*Part 63 subpart*	*APC review*[1,2]
Fuel combustion					
Engine test facilities	11/15/00				
Industrial boilers	11/15/00				
Institutional/commercial boilers	11/15/00				
Process heaters	11/15/00				
Stationary internal combustion engines	11/15/00				
Stationary turbines	11/15/00				
Liquid distribution					
Gasoline distribution (Stage 1)[6]	11/15/94	02/08/94	12/14/94	R	5:2, 2.20
Marine vessel loading operations[6]	11/15/97	05/13/94	09/19/95	Y	6:1, 2.1
Organic liquid distribution (nongasoline)	11/15/00				
Mineral product processing					
Alumina processing	11/15/00				
Asphalt/coal tar application—metal pipes	11/15/00				
Asphalt concrete manufacturing	11/15/00				
Asphalt processing	11/15/00				
Asphalt roofing manufacturing	11/15/00				
Chromium refractory production	11/15/00				
Clay product manufacturing	11/15/00				
Lime manufacturing	11/15/00				
Mineral wool production	11/15/97	05/08/97		DDD	7:4, 2.40
Portland cement manufacturing	11/15/97	03/24/98		LLL	8:4, 2.13
Taconite iron ore processing	11/15/00				
Wool fiberglass manufacturing	11/15/97	03/31/97		NNN	7:4, 2.16
Nonferrous metal processing					
Aluminum production, primary	11/15/97	09/26/96	10/07/97	LL	8:1, 2.15
Aluminum production, secondary	11/15/97				
Copper smelting, primary	11/15/97	04/20/98		QQQ	8:4, 2.34
Lead smelting, primary	11/15/97	04/17/98		TTT	8:4, 2.13
Lead smelting, secondary	11/15/94	06/09/94	06/23/95	X	5:5, 2.33
Magnesium refining, primary	11/15/00				
Petroleum and natural gas production and refining					
Oil and natural gas production	11/15/97	02/06/98		HH	8:3, 4.1
Natural gas transmission and storage	11/15/97	02/06/98		HHH	8:3, 4.1
Petroleum refineries—catalytic cracking (fluid and other) units, catalytic reforming units, and sulfur plant units	11/15/97	09/11/98		UUU	—[7]
Petroleum refineries—other sources not distinctly listed	11/15/94	07/15/94	08/18/95	CC	5:6, 4.1

TABLE 5-5
Status of MACT Standards—Continued

Industry group and source category	*Deadline*	*Date proposed*	*Date promulgated*	*Part 63 subpart*	*APC review*[1,2]
Pharmaceutical production processes					
Pharmaceutical production	11/15/97	04/02/97	09/21/98	GGG[8]	—[7]
Polymer and resin production					
Acetal resin production	11/15/97				
Acrylonitrile-butadiene-styrene resin production	11/15/94	03/29/95	09/12/96	JJJ[8]	7:1, 4.31
Alkyd resin production	11/15/00				
Amino resin production	11/15/97				
Boat manufacturing	11/15/00				
Butyl rubber production	11/15/94	06/12/95	09/05/96	U[8]	7:1, 4.3
Carboxymethylcellulose production	11/15/00				
Cellophane production	11/15/00				
Cellulose ether production	11/15/00				
Epichlorohydrin elastomer production	11/15/94	06/12/95	09/05/96	U[8]	7:1, 4.3
Epoxy resin production	11/15/94	05/16/94	03/08/95	W	5:3, 2.26
Ethylene-propylene rubber production	11/15/94	06/12/95	09/05/96	U[8]	7:1, 4.3
Flexible polyurethane foam production	11/15/97	12/27/96	10/07/98	III	—[7]
Hypalon™ production	11/15/94	06/12/95	09/05/96	U[8]	7:1, 4.3
Maleic anhydride copolymer production	11/15/00				
Methylcellulose production	11/15/00				
Methyl methacrylate-acrylonitrile-butadiene-styrene resin production	11/15/94	03/29/95	09/12/96	JJJ[8]	7:1, 4.31
Methyl methacrylate-butadiene-styrene terpolymer production	11/15/94	03/29/95	09/12/96	JJJ[8]	7:1, 4.31
Neoprene production	11/15/94	06/12/95	09/05/96	U[8]	7:1, 4.3
Nitrile butadiene rubber production	11/15/94	06/12/95	09/05/96	U[8]	7:1, 4.3
Nitrile resins	11/15/00	03/29/95	09/12/96	JJJ[8]	7:1, 4.31
Non-nylon polyamide production	11/15/94	05/16/94	03/08/95	W	5:3, 2.26
Phenolic resin production	11/15/97				
Polybutadiene rubber production	11/15/94	06/12/95	09/05/96	U[8]	7:1, 4.3
Polycarbonate production	11/15/97	12/31/92	04/22/94	H, I[4]	4:3, 4.1
Polyester resin production	11/15/00				
Polyether polyol production	11/15/97	09/04/97		PPP	7:6, 2.10
Polyethylene terephthalate resin production	11/15/94	03/29/95	09/12/96	JJJ[8]	7:1, 4.31
Polymerized vinylidene chloride production	11/15/00				
Polymethyl methacrylate resin production	11/15/00				
Polystyrene resin production	11/15/94	03/29/95	09/12/96	JJJ[8]	7:1, 4.31
Polysulfide rubber production	11/15/94	06/12/95	09/05/96	U[8]	7:1, 4.3
Polyvinyl acetate emulsion production	11/15/00				
Polyvinyl alcohol production	11/15/00				
Polyvinyl butyral production	11/15/00				
Polyvinyl chloride and copolymer production	11/15/00				
Reinforced plastic composites production	11/15/00				
Styrene-acrylonitrile resin production	11/15/94	03/29/95	09/12/96	JJJ[8]	7:1, 4.31
Styrene-butadiene rubber and latex production	11/15/94	06/12/95	09/05/96	U[8]	7:1, 4.3

TABLE 5-5
Status of MACT Standards—Continued

Industry group and source category	Deadline	Date proposed	Date promulgated	Part 63 subpart	APC review[1,2]
Production of inorganic chemicals					
Ammonium sulfate production—caprolactam by-product plants	11/15/00				
Antimony oxide manufacturing	11/15/00				
Carbon black production	11/15/00				
Chlorine production	11/15/00				
Cyanide chemical manufacturing	11/15/00				
Fume silica production	11/15/00				
Hydrochloric acid production	11/15/00				
Hydrogen fluoride production	11/15/00				
Phosphate fertilizer production	11/15/97	12/27/96		BB	7:2, 2.14
Phosphoric acid manufacturing	11/15/97	12/27/96		BB	7:2, 2.14
Uranium hexafluoride production	11/15/00				
Production of organic chemicals					
Ethylene processes	11/15/00				
Quaternary ammonium compound production	11/15/00				
Synthetic organic chemical manufacturing industry	11/15/92	12/31/92	04/22/94	F, G, H	4:3, 4.1
Surface coating processes					
Aerospace industries	11/15/94	06/06/94	09/01/95	GG	5:6, 2.14
Auto and light-duty truck (surface coating)	11/15/00				
Flat wood paneling (surface coating)	11/15/00				
Large appliance (surface coating)	11/15/00				
Magnetic tapes (surface coating)	11/15/94	03/11/94	12/15/94	EE	5:2, 2.46
Manufacture of paints, coatings, and adhesives	11/15/00				
Metal can (surface coating)	11/15/00				
Metal coil (surface coating)	11/15/00				
Metal furniture (surface coating)	11/15/00				
Miscellaneous metal parts and products (surface coating)	11/15/00				
Paper and other webs (surface coating)	11/15/00				
Plastic parts and products (surface coating)	11/15/00				
Printing, coating, and dyeing of fabrics	11/15/00				
Printing/publishing (surface coating)	11/15/94	03/14/95	05/30/96	KK	6:6, 2.3
Shipbuilding and ship repair (surface coating)	11/15/94	12/06/94	12/15/95	II	6:2, 2.29
Wood furniture (surface coating)	11/15/94	12/06/94	12/07/95	JJ	6:2, 2.2
Waste treatment and disposal					
Hazardous waste incineration	11/15/00	04/19/96		EEE	6:4, 2.2
Municipal landfills	11/15/00				
Offsite waste and recovery operations	11/15/94	10/13/94	07/01/96	DD, OO–RR, VV	6:6, 2.23

TABLE 5-5
Status of MACT Standards—Continued

Industry group and source category	*Deadline*	*Date proposed*	*Date promulgated*	*Part 63 subpart*	*APC review*[1,2]
Waste treatment and disposal—Continued					
Publicly owned treatment works emissions	11/15/95				
Sewage sludge incineration[9]	11/15/00				
Site remediation	11/15/00				
Miscellaneous processes					
Aerosol can-filling facilities	11/15/00				
Benzyltrimethylammonium chloride production	11/15/00				
Carbonyl sulfide production	11/15/00				
Chelating agent production	11/15/00				
Chlorinated paraffin production	11/15/00	12/31/92	04/22/94	H, I[4]	4:3, 4.1
Chromic acid anodizing	11/15/94	12/16/93	01/25/95	N	5:3, 2.2
Commercial sterilization facilities	11/15/94	03/07/94	12/06/94	O	5:2, 2.38
Decorative chromium electroplating	11/15/94	12/16/93	01/25/95	N	5:3, 2.2
Dry cleaning (perchloroethylene)—dry-to-dry and transfer machines	11/15/92	12/09/91	09/22/93	M	4:1, 2.22
Dry cleaners (petroleum solvent)	11/15/00				
Ethylidene norbornene production	11/15/00	12/31/92	04/22/94	H, I[4]	4:3, 4.1
Explosive production	11/15/00				
Flexible polyurethane foam fabrication operations	11/15/00				
Friction product manufacturing	11/15/00				
Halogenated solvent cleaners	11/15/94	11/29/93	12/02/94	T	5:2, 2.10
Hard chromium electroplating	11/15/94	12/16/93	01/25/95	N	5:3, 2.2
Hydrazine production	11/15/00				
Industrial process cooling towers	11/15/94	08/12/93	09/08/94	Q	5:1, 2.41
Leather tanning and finishing operations	11/15/00				
OBPA/1,3-diisocyanate production	11/15/00	12/31/92	04/22/94	H, I[4]	4:3, 4.1
Paint stripper users	11/15/00				
Photographic chemical production	11/15/00				
Phthalate plasticizer production	11/15/00	12/31/92	04/22/94	H, I[4]	4:3, 4.1
Plywood/particle board manufacturing	11/15/00				
Pulp and paper production					
—Kraft, soda, sulfite, semi-chemical, mechanical pulping; and non-wood fiber pulping	11/15/97	12/17/93	04/15/98	S	8:4, 4.1
—Chemical recovery combustion sources	11/15/00	04/15/98		MM	8:4, 4.1
Rocket engine test firing	11/15/00				
Rubber chemical manufacturing	11/15/00				
Semiconductor manufacturing	11/15/00				
Symmetrical tetrachloropyridine production	11/15/00	12/31/92	04/22/94	H, I[4]	4:3, 4.1
Tire production	11/15/00				

TABLE 5-5
Status of MACT Standards—Continued

Industry group and source category	Deadline	Date proposed	Date promulgated	Part 63 subpart	APC review[1,2]
Categories of area sources					
Chromic acid anodizing	11/15/94	12/16/93	01/25/95	N	5:3, 2.2
Commercial sterilization facilities	11/15/94	03/07/94	12/06/94	O	5:2, 2.38
Decorative chromium electroplating	11/15/94	12/16/93	01/25/95	N	5:3, 2.2
Dry cleaning (perchloroethylene)—dry-to-dry and transfer machines	11/15/92	12/09/91	09/22/93	M	4:1, 2.22
Halogenated solvent cleaners	11/15/94	11/29/93	12/02/94	T	5:2, 2.10
Hard chromium electroplating	11/15/94	12/16/93	01/25/95	N	5:3, 2.2
Lead smelting, secondary	11/15/97	06/09/94	06/23/95	X	5:5, 2.33
Oil and natural gas production[10]	11/15/97	02/06/98		HH	8:3, 4.1

APC = *The Air Pollution Consultant*; HAP = hazardous air pollutant; HCl = hydrochloric acid; HON = hazardous organic NESHAP; LDAR = leak detection and repair; MACT = maximum achievable control technology; NESHAP = national emission standards for hazardous air pollutants; OBPA = oxybisphenoxarsine.

[1]Where a particular standard has been proposed *and* promulgated, the reference applies to the final rule; where the standard has only been proposed, the reference applies to the proposed rule. Proposed and final *revisions* to the original rule are *not* included in this table.

[2]Reference indicates the volume and issue in which an article reviewing the particular rule appeared in *The Air Pollution Consultant,* for example 8:2, 2.6 means the article appears in Volume 8, Issue 2, beginning on page 2.6.

[3]In the November 10, 1997 preamble to the proposed Subpart MMM MACT standards, EPA also noted its plan to combine the 11 individual agricultural chemical sources categories into a single category under this title.

[4]Subparts H and I are the LDAR provisions of the HON rule; these provisions apply to the emission of certain chemicals from facilities in this source category (see §63.190 for details).

[5]This rule was reproposed on December 4, 1992.

[6]This source category has been incorporated into the petroleum refinery source category for purposes of emission averaging only. This category is otherwise regulated under the rule listed in this entry.

[7]This MACT standard will be reviewed in the January/February 1999 issue of *APC.*

[8]HON rule LDAR provisions of 40 *CFR* Part 63, Subparts H and I apply to these sources until the source category-specific MACT standard compliance deadline.

[9]Although EPA stated on January 14, 1997 (62 *FR* 1868) that it would issue MACT for this source category under CAA Section 129, it has not yet removed this listing from this (Section 112) list.

[10]EPA noted its plans to add this source category when it proposed the listed Subpart HH MACT standards.

Source: Compiled by Elsevier Science Inc.

TABLE 5-6
Recommended Sampling and Analytical Methods for Hazardous Air Pollutants Regulated Under the NESHAP Program

CAS no.	Chemical name	Sampling method	Analytical method
75-07-0	Acetaldehyde	0011	0011A, 8315
60-35-5	Acetamide	0010	8270
75-05-8	Acetonitrile	0030	5040, 5041
98-86-2	Acetophenone	0010	8270
53-96-3	2-Acetylamino-fluorene	0010	8270
107-02-8	Acrolein	0011	8315
		0030	5040, 5041
79-06-1	Acrylamide	0030	5040, 5041
79-10-7	Acrylic acid	0010	8270
107-13-1	Acrylonitrile	0030	5040, 5041
107-05-1	Allyl chloride	0030	5040, 5041[a]
92-67-1	4-Aminobiphenyl	0010	8270
62-53-3	Aniline	0010	8270
90-04-0	*o*-Anisidine	0010	8270
1332-21-4	Asbestos	CARB 427	—
71-43-2	Benzene (including benzene from gasoline)	0030	5040, 5041
92-87-5	Benzidine	0010	8270
98-07-7	Benzotrichloride	0010	8270
100-44-7	Benzyl chloride	0010	8270
92-52-4	Biphenyl	0010	8270, 8310
542-88-1	Bis(chloromethyl) ether	0030	5040, 5041[a]
117-81-7	Bis(2-ethylhexyl) phthalate (DEHP)	0010	8270
75-25-2	Bromoform	0010	8270
106-99-0	1,3-Butadiene	18	GC/FID, GC/MS
156-62-7	Calcium cyanamide	None	—
133-06-2	Captan	0010	8270
63-25-2	Carbaryl	0010	8313, HPLC
75-15-0	Carbon disulfide	0030	5040, 5041
56-23-5	Carbon tetrachlo-ride	0030	5040, 5041
463-58-1	Carbonyl sulfide	15	—
120-80-9	Catechol	0010	8270
133-90-4	Chloramben	0010	HPLC
57-74-9	Chlordane	0010	8270, 8080
7782-50-5	Chlorine	0050, 0051	9057
79-11-8	Chloroacetic acid	0010	8270
532-27-4	2-Chloroaceto-phenone	0010	8270, 8315
		0011	8315
108-90-7	Chlorobenzene	0010	8270
		0030	5040, 5041
510-15-6	Chlorobenzilate	0010	8270[a]
67-66-3	Chloroform	0030	5040, 5041
107-30-2	Chloromethyl methyl ether	0030	5040, 5041
126-99-8	Chloroprene	0030	5040, 5041[a]
1319-77-3	Cresols/cresylic acid (isomers and mixtures)	0010	8270
108-39-4	*m*-Cresol	0010	8270
95-48-7	*o*-Cresol	0010	8270
106-44-5	*p*-Cresol	0010	8270
98-82-8	Cumene	0010	8270
94-75-7	2,4-D, salts and esters	0010	8270
3547-04-4	DDE	0010	8270
334-88-3	Diazomethane	None	—
132-64-9	Dibenzofurans	0010	8270
96-12-8	1,2-Dibromo-3-chloropropane	0010	8270
84-74-2	Dibutylphthalate	0010	8270
106-46-7	1,4-Dichloro-benzene(p)	0010	8270
91-94-1	3,3-Dichloro-benzidene	0010	8270[a]
111-44-4	Dichloroethyl ether [Bis(2-chloro-ethyl)ether]	0010	8270
542-75-6	1,3-Dichloro-propene	0010	8270
		0030	5040, 5041
62-73-7	Dichlorvos	0010	8141
111-42-2	Diethanolamine	0010	8270
121-69-7	N,N-Diethylani-line (N,N-Dimethylaniline)	0010	8270
64-67-5	Diethyl sulfate	0010	8270
119-90-4	3,3-Dimethoxy-benzidine	0010	8270
60-11-7	Dimethyl amino-azobenzene	0010	8270
119-93-7	3,3′-Dimethyl benzidine	0010	8270

TABLE 5-6
Recommended Sampling and Analytical Methods for Hazardous Air Pollutants Regulated Under the NESHAP Program—Continued

CAS no.	Chemical name	Sampling method	Analytical method
79-44-7	Dimethyl carbamoyl chloride	0010	8270
68-12-2	Dimethyl formamide	0010	8270
57-14-7	1,1-Dimethyl hydrazine	0030	5040, 5041
131-11-3	Dimethyl phthalate	0010	8270
77-78-1	Dimethyl sulfate	0010	8270
534-52-1	4,6-Dinitro-*o*-cresol and salts	0010	8270
51-28-5	2,4-Dinitrophenol	0010	8270
121-14-2	2,4-Dinitro-toluene	0010	8270
123-91-1	1,4-Dioxane (1,4-Diethylene-oxide)	0010	8270
122-66-7	1,2-Diphenyl-hydrazine	0010	8270
106-89-8	Epichlorohydrin (l-Chloro-2,3-epoxypropane)	0010 0030	8270[a] 5040, 5041
106-88-7	1,2-Epoxybutane	0030	5040, 5041
140-88-5	Ethyl acrylate	0010	8270
100-41-4	Ethyl benzene	0010	8270
51-79-6	Ethyl carbamate (Urethane)	0010	8270
75-00-3	Ethyl chloride (Chloroethane)	0030	5040, 5041
106-93-4	Ethylene dibromide (Dibromo-ethane)	0030	5040, 5041[a]
107-06-2	Ethylene dichloride (1,2-Dichloro-ethane)	0030	5040, 5041
107-21-1	Ethylene glycol	0010	8270
151-56-4	Ethylene imine (Aziridine)	0030	5040, 5041
75-21-8	Ethylene oxide	18	GC/FID
96-45-7	Ethylene thiourea	0010	632
75-34-3	Ethylidene di-chloride (1,1-Dichloroethane)	0030	5040, 5041
50-00-0	Formaldehyde	0011	0011A, 8315
76-44-8	Heptachlor	0010	8270
118-74-1	Hexachlorobenzene	0010	8270
87-68-3	Hexachloro-butadiene	0010	8270
77-47-4	Hexachlorocyclo-pentadiene	0010	8270[a]
67-72-1	Hexachloroethane	0010	8270
822-06-0	Hexamethylene-1,6-diisocyanate	0010	8270
680-31-9	Hexamethyl-phoshoramide	0010	8270
110-54-3	Hexane	0030	5040, 5041
302-01-2	Hydrazine	0010 0030	8270 5040, 5041
7647-01-0	Hydrochloric acid	0050, 0051, 26	9057
7664-39-3	Hydrogen fluoride (Hydrofluoric acid)	13A, 13B	Ion chroma-tography
123-31-9	Hydroquinone	0010 0011	8270 8315
78-59-1	Isophorone	0010	8270
58-89-9	Lindane (all isomers)	0010	8080, 8270
108-31-6	Maleic anhydride	0010	8270
67-56-1	Methanol	18	18
72-43-5	Methoxychlor	0010	8270
74-83-9	Methyl bromide (Bromomethane)	0030	5040, 5041
74-87-3	Methyl chloride (Chloromethane)	0030 18	5040, 5041[a] 5040, 5041
71-55-6	Methyl chloro-form (1,1,1-Trichloro-ethane)	0030	5040, 5041
78-93-3	Methyl ethyl ketone (2-Butanone)	0011 0030	8315 5040, 5041
60-34-4	Methyl hydrazine	0030	5040, 5041
74-88-4	Methyl iodide (Iodomethane)	0030	5040, 5041

TABLE 5-6
Recommended Sampling and Analytical Methods for Hazardous Air Pollutants Regulated Under the NESHAP Program—Continued

CAS no.	Chemical name	Sampling method	Analytical method
108-10-1	Methyl isobutyl ketone (Hexone)	0010 0011	8270 8315
624-83-9	Methyl isocyanate	0030	5040, 5041
80-62-6	Methyl methacrylate	0030	5040, 5041
1634-04-4	Methyl *tert*-butyl ether	0030	5040, 5041
101-14-4	4,4-Methylene bis (2-Chloro-aniline)	0010	8270
75-09-2	Methylene chloride (Di-chloromethane)	0030	5040, 5041
101-68-8	Methylene diphenyl diisocyanate (MDI)	0010	8270, HPLC
101-77-9	4,4′-Methylene-dianiline	0010	8270
91-20-3	Naphthalene	0010	8270, 8310
98-95-3	Nitrobenzene	0010	8270
92-93-3	4-Nitrobiphenyl	0010	8270
100-02-7	4-Nitrophenol	0010	8270
79-46-9	2-Nitropropane	0010 0030	8270 5040, 5041
684-93-5	N-Nitroso-N-methylurea	0010	8270
62-75-9	N-Nitrosodi-methylamine	0010	8270
59-89-2	N-Nitrosomorpho-line	0010	8270
56-38-2	Parathion	0010	8270
82-68-8	Pentachloronitro-benzene (Quintobenzene)	0010	8270
87-86-5	Pentachlorophenol	0010	8270
108-95-2	Phenol	0010	8270
106-50-3	*p*-Phenylene-diamine	0010	8270
75-44-5	Phosgene	Impinger derivative solution	HPLC/UV, GC/FID, GC/MS
7803-51-2	Phosphine	0012	GC/FPD, 0012
7723-14-0	Phosphorus	0012	0012
85-44-9	Phthalic anhydride	0010	8270
1336-36-3	Polychlorinated biphenyls (Aroclors)	0010	8080, 8270
1120-71-4	1,3-Propane sultone	0010	8270
57-57-8	*beta*-Propiolactone	0010	8270
123-38-6	Propionaldehyde	0011	8315
114-26-1	Propoxur (Baygon)	0010	8318
78-87-5	Propylene dichloride (1,2-Dichloro-propane)	0030	5040, 5041
75-56-9	Propylene oxide	0030 18	5040, 5041
75-55-8	1,2-Propylenimine (2-Methyl aziridine)	0030	5040, 5041
91-22-5	Quinoline	0010	8270
106-51-4	Quinone	0010 0011	HPLC/Electro-chemistry 8315
100-42-5	Styrene	0010	8270
96-09-3	Styrene oxide	0010	8270
1746-01-6	2,3,7,8-Tetrachloro-dibenzo-*p*-dioxin	0010 23	8280, 8290 —
79-34-5	1,1,2,2-Tetra-chloroethane	0010	8270
127-18-4	Tetrachloroethylene (Perchloro-ethylene)	0010 0030	8270 5040, 5041
7550-45-0	Titanium tetrachloride	0012	0012
108-88-3	Toluene	0010 0030	8270 5040, 5041
95-80-7	2,4-Toluene diamine	0010	8270
584-84-9	2,4-Toluene diisocyanate	Impinger	HPLC, UV
95-53-4	*o*-Toluidine	0010	8270
8001-35-2	Toxaphene (chlorinated camphene)	0010	8270, 8250, 8080
120-82-1	1,2,4-Trichloroben-zene	0010	8270, 8250

TABLE 5-6
Recommended Sampling and Analytical Methods for Hazardous Air Pollutants Regulated Under the NESHAP Program—Continued

CAS no.	Chemical name	Sampling method	Analytical method
79-00-5	1,1,2-Trichloroethane	0010 0030	8270 5040, 5041
79-01-6	Trichloroethylene	0030	5040, 5041
95-95-4	2,4,5-Trichlorophenol	0010	8270
88-06-2	2,4,6-Trichlorophenol	0010	8270
121-44-8	Triethylamine	0030	5040, 5041
1582-09-8	Trifluralin	0010	8270
540-84-1	2,2,4-Trimethylpentane	0030	5040, 5041
108-05-4	Vinyl acetate	0030	5040, 5041
593-60-2	Vinyl bromide	106	GC/MS
75-01-4	Vinyl chloride	0030 106	5040, 5041 5040, 5041, GC/MS, 106
75-35-4	Vinylidene chloride (1,1-Dichloroethylene)	0030	5040, 5041
1330-20-7	Xylenes (isomers and mixtures)	0010	8270
108-38-3	*m*-Xylenes	0010	8270
95-47-6	*o*-Xylenes	0010	8270
106-42-3	*p*-Xylenes	0010	8270
—	Antimony compounds	0012	0012
—	Arsenic compounds (inorganic including arsine)	0012	0012
—	Beryllium compounds	0012	0012
—	Cadmium compounds	0012	0012
—	Chromium compounds	0012	0012
—	Cobalt compounds	0012	0012
—	Coke-oven emissions	0010	8310
—	Cyanide compounds	Filter, Impinger	Extract, SEI
—	Glycol ethers	0010	8270
—	Lead compounds	0012	EPA 12, 0012
—	Manganese compounds	0012	6010
—	Mercury compounds	0012	EPA 101, 0012
—	Fine mineral fibers	CARB 427	—
—	Nickel compounds	0012	0012
—	Polycylic organic matter	CARB 429	—
—	Radionuclides (including radon)	EPA 111, 114, 115	—
—	Selenium compounds	0012	0012

CARB = California Air Resources Board; CAS = Chemical Abstract Service; DDE = 1,1-dichloro-2,2-bis(*p*-chlorophenylethylene); GC/FID = gas chromatography/flame ionization detector; GC/FPD = gas chromatography/flame photometric detector; GC/MS = gas chromatography/mass spectrometry; HPLC = high-pressure liquid chromatography; HPLC/UV = high-pressure liquid chromatography/ultraviolet absorption; NESHAP = national emission standards for hazardous air pollutants; UV = ultraviolet absorption.

[a]These methods have not met Method 301 validation criteria.

Source: Adapted from L.E. Cottone et al., TRC/North American Weather Consultants (Salt Lake City, Utah), "A Summary of Emission Measurement Methods for Title III Clean Air Act Amendment HAPs," presented at the 1994 Annual Air & Waste Management Meeting.

TABLE 5-7
NESHAP Program Regulatory Compliance Calendar[1]

Compliance date[2]: Existing sources	Compliance date[2]: New sources	New source threshold date[3]	Applicability and regulatory requirements[2]	CFR reference (Title 40)	FR reference	APC review[4]
General NESHAP provisions						
Specified in the applicable MACT standard	Same as existing sources	Specified in the applicable MACT standard	Sources subject to a MACT standard must also implement the general NESHAP provisions spelled out in Part 63, Subpart A. Applicable general provisions are incorporated by reference within individual source category-specific MACT standards; these typically include performance testing, monitoring, recordkeeping, and reporting.	Part 63, Subpart A	03/16/94 59 *FR* 12408	4:4, 4.2
Section 112(g) preconstruction review program						
NA	At start-up	01/27/97	Newly constructed and reconstructed major HAP sources in source categories for which MACT has not yet been promulgated must define case-by-case MACT and implement controls at start-up. Prior to construction, regulated sources must also 1) obtain or revise their Title V permit, or 2) conduct a preconstruction review per the terms of the Section 112(g) rule or an approved state or local program. New HAP sources connected to certain existing control equipment are exempt from the program.	Part 63, Subpart B	12/27/96 61 *FR* 68384	7:2, 2.2
Dry cleaning MACT						
12/21/93 (LDAR) 09/22/96 (other provisions)	Start-up or 09/22/93, whichever is later (all provisions)	12/09/91	Industrial and commercial dry cleaning facilities that use perchloroethylene must implement emission controls and an LDAR program; level of control specified varies depending on whether the facility is a major HAP source and the type of dry cleaning equipment in use.	Part 63, Subpart M	09/22/93 58 *FR* 49354 12/20/93 58 *FR* 66287	4:1, 2.22 (original rule) 4:2, 2.62 (minor rule changes)
			Transfer-type dry cleaning machines installed between December 9, 1991 and September 22, 1993 may operate subject to certain controls (such units were banned under the original rule).		09/19/96 61 *FR* 49263	7:1, 2.39 (ban rescission)

TABLE 5-7
NESHAP Program Regulatory Compliance Calendar[1]—Continued

Compliance date[2]		*New source threshold date*[3]	*Applicability and regulatory requirements*[2]	*CFR reference (Title 40)*	*FR reference*	*APC review*[4]
Existing sources	*New sources*					
HAP early reduction compliance extensions						
Proposal date of applicable MACT standards or 01/01/94, whichever is later	Same as existing sources	NA	Sources that reduce HAP emissions by 90% (95% for particulate HAP) via an alternative emission limitation that is documented in a Title V operating permit may obtain a six-year extension from an applicable MACT standard compliance deadline.	Part 63, Subpart D	12/29/92 57 *FR* 61970	3:2, 2.34[5]
HON rule						
Various dates beginning 10/24/94[6,7] (LDAR)	Start-up or 04/22/94, whichever is later (all provisions)[6]	12/31/92	Major HAP sources must implement MACT for chemical process units that manufacture a SOCMI chemical as their primary product. Regulated emission points include process vents, storage vessels, transfer racks, wastewater handling and treatment facilities, and fugitive equipment leaks.	Part 63, Subparts F–H	04/22/94 59 *FR* 19402	4:3, 4.1 and 4:4, 2.25[5]
04/22/97 (all other provisions)			Major HAP source production units belonging to certain non-SOCMI source categories listed in §63.190(b) must control fugitive equipment leaks (i.e., LDAR).	Part 63, Subpart I		
Coke oven battery MACT						
12/31/95[8]	Start-up or 10/27/93, whichever is later	11/15/90[9]	All by-product recovery and recovery coke oven batteries must implement MACT.	Part 63, Subpart L	10/27/93 58 *FR* 57898	4:1, 2.29 (original rule)
					01/13/94 59 *FR* 1992	4:3, 2.34 (corrections)

TABLE 5-7

NESHAP Program Regulatory Compliance Calendar[1]—Continued

Compliance date[2]		*New source threshold date[3]*	*Applicability and regulatory requirements[2]*	*CFR reference (Title 40)*	*FR reference*	*APC review[4]*
Existing sources	*New sources*					
Chromium electroplating and anodizing MACT						
01/25/96 (decorative chromium electroplating) 01/25/97 (hard chromium electroplating and chromium anodizing)	Start-up or 01/25/95, whichever is later (all sources)	12/16/93	All sources must implement MACT for decorative chromium electroplating, hard chromium electroplating, and chromium anodizing facilities.	Part 63, Subpart N	01/25/95 60 *FR* 4948 01/30/97 62 *FR* 4463	5:3; 2.2 7:3, 2.15 (California hard electroplating and anodizing source monitoring extension)
Industrial process cooling tower MACT						
03/08/96	Start-up or 09/08/94, whichever is later	08/12/93	Major HAP sources must discontinue the use of chromium-based water treatment chemicals in their industrial process cooling towers.	Part 63, Subpart Q	09/08/94 59 *FR* 46339	5:1; 2.41[5]
Magnetic tape manufacturing MACT						
12/15/96 (if new add-on control device not needed to comply) 12/15/97 (if new add-on control device needed to comply)	Start-up or 12/15/94, whichever is later	03/11/94	Major HAP sources must implement MACT for magnetic tape manufacturing operations.	Part 63, Subpart EE	12/15/94 59 *FR* 64580	5:2, 2.46

TABLE 5-7
NESHAP Program Regulatory Compliance Calendar[1]—Continued

Compliance date[2]		New source threshold date[3]	Applicability and regulatory requirements[2]	CFR reference (Title 40)	FR reference	APC review[4]
Existing sources	New sources					
Elastomer production MACT						
07/31/97 (LDAR)[6] 09/05/99 (other provisions)	Start-up or 9/05/96, whichever is later[6]	06/12/95	Major HAP sources must implement MACT for elastomer production units; regulated emission points include storage vessels, process vents, heat exchange systems, process contact cooling towers, wastewater facilities, and fugitive equipment leaks.	Part 63, Subpart U	09/05/96 61 *FR* 46906 01/14/97 62 *FR* 1835	7:1, 4.3 7:2, 2.54 (LDAR compliance extension)
Wood furniture MACT						
11/21/97 (facilities emitting ≥50 tpy HAPs) 12/07/98 (other sources)	Start-up or 12/07/95, whichever is later	12/06/94	Major HAP sources must implement MACT for wood furniture manufacturing facilities.	Part 63, Subpart JJ	12/07/95 60 *FR* 62930	6:2, 2.2[5]
Halogenated solvent cleaning operations MACT						
12/02/97 08/03/98 (continuous web cleaning machines)	Start-up or 12/02/94, whichever is later	11/29/93	All facilities with batch or in-line solvent cleaning machines using ≥5-wt% methylene chloride, perchloroethylene, trichloroethylene, 1,1,1-trichloroethane, or carbon tetrachloride must implement MACT for those machines.	Part 63, Subpart T	12/02/94 59 *FR* 61801 12/30/94 59 *FR* 67750	5:2; 2.10 (original rule) 5:2, 2.71 (compliance date correction)
Gasoline distribution facility MACT						
12/15/97	Start-up or 12/14/94, whichever is later	02/08/94	Major HAP sources must implement MACT for gasoline distribution facilities.	Part 63, Subpart R	12/14/94 59 *FR* 64303 02/29/96 61 *FR* 7718	5:2; 2.20 (original rule) 6:3, 2.14 (LDAR compliance extension)

TABLE 5-7

NESHAP Program Regulatory Compliance Calendar[1]—Continued

Compliance date[2]		*New source threshold date[3]*	*Applicability and regulatory requirements[2]*	*CFR reference (Title 40)*	*FR reference*	*APC review[4]*
Existing sources	*New sources*					
Secondary lead smelter MACT						
12/23/97	Start-up or 06/23/95, whichever is later	06/09/94	All sources must implement MACT for secondary lead smelters.	Part 63, Subpart X	06/23/95 60 *FR* 32587	5:5, 2.33
					12/12/96 61 *FR* 65334	7:2, 2.58 (compliance extension)
Shipbuilding and ship repair MACT						
12/16/97	Start-up or 12/16/96, whichever is later	12/06/94	Major HAP sources must implement MACT for shipbuilding and ship repair facilities.	Part 63, Subpart II	12/15/95 60 *FR* 64330	6:2, 2.29 (original rule)
					06/18/96 61 *FR* 30814	6:6, 2.35 (compliance extension)
Thermoplastic resin MACT						
02/27/98 (LDAR)[6]	Start-up or 02/27/98, whichever is later[6]	06/29/95	Major HAP sources must implement MACT for thermoplastic resin production units; regulated emission points include storage vessels, process vents, process back-end operations, wastewater facilities, and fugitive equipment leaks.	Part 63, Subpart JJJ	09/12/96 61 *FR* 48208	7:1, 4.31[5]
09/12/99 (other provisions)					01/14/97 62 *FR* 1835	7:2, 2.54 (LDAR compliance extension)
					02/27/98 63 *FR* 9944	8:3, 2.28 (overall additional compliance extension)

TABLE 5-7

NESHAP Program Regulatory Compliance Calendar[1]—Continued

Compliance date[2]		*New source threshold date[3]*	*Applicability and regulatory requirements[2]*	*CFR reference (Title 40)*	*FR reference*	*APC review[4]*
Existing sources	*New sources*					
Epichlorohydrin-based resin MACT						
03/08/98	Start-up or 03/08/95, whichever is later	05/16/94	Major HAP sources must implement MACT for basic liquid epoxy and wet-strength resin production units.	Part 63, Subpart W	03/08/95 60 *FR* 12670	5:3; 2.26
Petroleum refinery MACT						
08/18/98[6]	Start-up or 08/18/95, whichever is later[6]	07/14/94	Major HAP sources must implement MACT for petroleum refinery process units.	Part 63, Subpart CC	08/18/95 60 *FR* 43244	5:6, 4.1[5]
Aerospace MACT						
09/01/98	Start-up or 09/01/95, whichever is later	06/06/94	Major HAP sources must implement MACT for facilities that manufacture or rework commercial, civil, or military aerospace vehicles or components.	Part 63, Subpart GG	09/01/95 60 *FR* 45948	5:6, 2.14[5]
Emission standards for marine vessel operations						
09/19/98[10]	Upon start-up	09/21/98[11]	Facilities whose marine vessel operation throughput is ≥10 million barrels/year of gasoline or 200 million barrels/year of crude oil must implement RACT for those operations.	Part 63, Subpart Y	09/19/95 60 *FR* 48388	6:1, 2.1
09/19/99	Start-up or 09/19/99, whichever is later	05/13/94	Existing sources whose marine vessel operations exceed the HAP major source threshold and *all* new sources must implement MACT for those operations.			

TABLE 5-7

NESHAP Program Regulatory Compliance Calendar[1]—Continued

Compliance date[2]		*New source threshold date[3]*	*Applicability and regulatory requirements[2]*	*CFR reference (Title 40)*	*FR reference*	*APC review[4]*
Existing sources	*New sources*					
Commercial (ethylene oxide) sterilization facility MACT						
12/06/98	Upon start-up	12/08/97[12]	All sources with commercial sterilization and fumigation operations that use more than 1 tpy of ethylene oxide must implement MACT for those facilities.	Part 63, Subpart O	12/06/94 59 *FR* 62585 12/09/97 62 *FR* 64736	5:2; 2.38 8:2, 2.31 (compliance extension)
Printing and publishing industry MACT						
05/30/99	Start-up or 05/30/96, whichever is later	03/14/95	Major HAP sources must implement MACT for printing/publishing facilities.	Part 63, Subpart KK	05/30/96 61 *FR* 27132	6:6, 2.3
Accidental release prevention program						
The later of 1) the date a substance is first handled above its TQ, 2) three years after EPA lists a new regulated substance, or 3) 06/21/99	Same as existing sources	NA	Facilities having regulated substances present in greater than the applicable TQ must implement a risk management plan; level of regulation depends on relative risks posed by the facility's processes.	Part 68, Subpart A	01/31/94 59 *FR* 4478	4:3, 2.17 (regulated substance list)[5]
Offsite waste and recovery operation MACT						
07/01/99	Start-up or 07/01/96, whichever is later	10/13/94	Major HAP sources must implement MACT for offsite material management units, process vents, and equipment leaks.	Part 63, Subparts DD, OO–RR, and VV	07/01/96 61 *FR* 34140	6:7, 2.23

TABLE 5-7
NESHAP Program Regulatory Compliance Calendar[1]—Continued

Compliance date[2]: Existing sources	Compliance date[2]: New sources	New source threshold date[3]	Applicability and regulatory requirements[2]	CFR reference (Title 40)	FR reference	APC review[4]
Primary aluminum reduction MACT						
10/07/99	Start-up or 10/07/97, whichever is later	09/26/96	Major HAP sources must implement MACT for their potlines, anode bake furnaces, paste production plants, and pitch storage tanks.	Part 63, Subpart LL	10/07/97 62 *FR* 16088	8:1, 2.15
Pulp and paper noncombustion source MACT						
04/16/01	Start-up or 06/18/93, whichever is later	12/17/98 (kraft, soda, sulfite, semi-chemical pulping processes) 03/08/96 (mechanical wood fiber pulping and nonwood fiber pulping)	Major HAP sources must implement MACT for their kraft, soda, sulfite, semi-chemical pulping processes; mechanical wood fiber pulping; and non-wood fiber pulping; in general, sources may 1) enclose, capture, and destroy, recover or otherwise reduce their HAP emissions; or 2) limit their HAP emissions based on overdried pulp production	Part 63 Subpart S	04/15/98 63 *FR* 18504	8:4, 4.1[5]
Pharmaceutical MACT						
09/21/01	Start-up or 09/21/98, whichever is later	04/02/97	Major HAP sources must implement MACT for their processes and associated equipment that produce pharmaceutical products; regulated emission points include storage tanks, process vents, wastewater treatment systems, heat exchange systems, and fugitive equipment leaks	Part 63 Subpart GGG	09/21/98 63 *FR* 50280	9:1, 4.1

TABLE 5-7

NESHAP Program Regulatory Compliance Calendar[1]—Continued

Compliance date[2]		*New source threshold date*[3]	*Applicability and regulatory requirements*[2]	*CFR reference (Title 40)*	*FR reference*	*APC review*[4]
Existing sources	*New sources*					
Flexible polyurethane foam production MACT						
10/08/01	Start-up or 10/08/98, whichever is later	12/27/96	Major HAP sources must implement MACT for their flexible polyurethane foam and rebond foam production units; regulated emission points include storage tanks, process vents, and fugitive equipment leaks	Part 63 Subpart III	10/08/98 63 *FR* 58980	—[13]

APC = *The Air Pollution Consultant*; HAP = hazardous air pollutant; HON= hazardous organic NESHAP; LDAR = leak detection and repair; MACT = maximum achievable control technology; NA = not applicable; NESHAP = national emission standards for hazardous air pollutants; RACT = reasonably available control technology; SOCMI = synthetic organic chemical manufacturing industry; TQ= threshold quantity.

[1]The calendar does *not* include rules and programs in the following five areas: 1) programs that are implemented on a state or local level where compliance deadlines are based on the individual agency's action (e.g., Title V operating permit programs and RACT requirements for sources in ozone nonattainment areas); 2) rules that impose deadlines on state or local government agencies (e.g., 40 *CFR* Part 58 regulations that require states and localities to establish ambient air quality monitoring systems); 3) rules promulgated prior to May 8, 1991; 4) rules that affect mobile sources; and 5) proposed rules that have not yet been promulgated.

[2]Various exemptions, extensions, and/or exceptions apply for some rules. These are *not* listed in this table.

[3]New sources are those for which construction or reconstruction began after the listed threshold date.

[4]Refers to the volume, issue, and page number of the applicable *APC* review article. For example, 4:4, 4.2 refers to the *APC* article beginning on page 4.2 of Volume 4, Issue 4.

[5]Other rulemakings affecting this program have been proposed or finalized since the original rule; these are *not* listed in the table. Refer to the *APC* cumulative index for a listing of review articles for these actions.

[6]Compliance with LDAR provisions occurs in progressively more stringent phases *beginning* on date listed.

[7]Compliance date for existing sources depends on "risk group" of SOCMI chemicals handled (risk group is assigned based on health hazard and carcinogenicity of chemical). Compliance dates are as follows: 1) Group I, October 24, 1994; 2) Group II, January 23, 1995; 3) Group III, April 24, 1995; 4) Group IV, July 24, 1995; and 5) Group V, October 23, 1995. See 40 *CFR* Part 63, Subpart F, Table 1 for SOCMI chemical risk group listings.

[8]More stringent emission limits take effect for existing sources on January 1, 2003. Existing sources may also choose alternate "early reduction" limits. These limits take effect on November 15, 1993 and increase in stringency on January 1, 1998 and again on January 1, 2010.

[9]New facilities are those that increase the design capacity of a coke oven battery above the unit's capacity as of November 15, 1990.

[10]Source must also provide EPA proof that it started construction of its vapor collection system and air pollution control device by September 19, 1997.

[11]Threshold is for compliance date determination only. For other aspects of the marine vessel RACT, a new source threshold is not meaningful.

[12]Threshold is for purposes of compliance date determination only. For all other aspects of the commercial sterilization MACT standards, a new source is one for which construction, reconstruction, or modification began after March 7, 1994.

[13]This standard will be reviewed in the January/February 1999 issue of the *APC*.

Source: Elsevier Science Inc.

The General NESHAP Provisions

Subpart A of Part 63 deals with certain "generic" regulatory requirements such as performance testing, monitoring, recordkeeping, and reporting. Rather than repeat common provisions in every MACT standard, EPA has spelled out these provisions in the so-called general NESHAP provisions (of Subpart A), which are then incorporated by reference into individual MACT standards. Table 5-8 (pages 5.27–5.31) is a quick reference to the requirements of Part 63, Subpart A.

Key general NESHAP provision requirements include:

- Compliance deadlines, which are listed in Table 5-9 (pages 5.32–5.33);
- Performance tests;
- Requirements for continuous emission monitoring, continuous opacity monitoring, and parameter monitoring;
- Start-up, shutdown, and malfunction plans; and
- Reporting and recordkeeping requirements. (Reporting requirements specified under the Part 63 general provisions are summarized in Table 5-10 [page 5.34].)

General NESHAP Performance Test Schedules

Facility owners/operators regulated under MACT standards must conduct performance tests of their air pollution control systems soon after the affected source must comply with the applicable standard (and when requested to do so by EPA or the state agency). The purpose of performance tests is to 1) establish whether the source is in compliance with applicable emission limits, and 2) establish operating parameter ranges to be used in ongoing compliance monitoring.

Stringent testing requirements have been included in the Part 63 general provisions to ensure that data obtained during performance tests are true indicators of a source's compliance status. Schedules for conducting performance tests at affected sources (if required) are set forth in §63.7 and are summarized in Table 5-11 (page 5.35).

For More Information

Refer to page 4.1 of the July/August 1994 issue of *APC* for a detailed article on the general NESHAP provisions.

The Accidental Release Prevention Program

The accidental release prevention program, established by the 1990 amendments, requires facilities to 1) take preventive action to avoid accidental chemical releases, and 2) prepare to mitigate the consequences of such accidents should they occur. Regulations implementing the program were promulgated on June 20, 1996 (61 *FR* 31668) and are codified at 40 *CFR* Part 68. The rule applies to every process that handles a regulated substance in quantities above the applicable threshold quantity (TQ) for that substance. Regulated substances consist of 1) the toxic substances listed in Table 5-12 (pages 5.36 and 5.37), and 2) the flammable substances listed in Table 5-13 (page 5.38). While a uniform TQ of 10,000 lbs applies to all flammables, toxic compound TQs vary from 500 to 20,000 lbs (as shown in Table 5-12).

Originally, it was intended that the accidental release prevention program also apply to certain explosive substances. However, because EPA determined that existing regulations issued by other federal agencies adequately control these substances, the agency removed explosives from the Part 68 program on January 6, 1998 (63 *FR* 640).

Stationary sources regulated under the accidental release prevention program must prepare risk management plans (RMPs) for their processes that handle a regulated substance in quantities above the applicable TQ. The scope of the required RMP is determined based on which program level the process fits into: Program 1, 2, or 3. The criteria for each of these programs, which depend on the likelihood and severity of potential accidental releases at a given facility, are summarized in Table 5-14 (page 5.39). Finally, Figure 5-1 (page 5.40) can be used to determine which program (and consequently which level of RMP) is required for a specific facility.

For further details on the original accidental release prevention program, please refer to the September/October 1996 issue of *APC*, page 4.1. Details on the list of substances regulated under Part 68 can be found on page 2.37 of the July/August 1996 issue of *APC* and page 2.17 of the May/June 1994 issue of *APC*.

Finally, readers should also see the *APC* cumulative keyword index for references to several other articles reviewing changes to the accidental release prevention program (e.g., a review of the deletion of explosives from the list of regulated substances can be found on page 2.20 of the May/June 1998 *APC*).

TABLE 5-8

Frequently Cited Part 63, Subpart A General NESHAP Regulations

Regulatory citation	*Description*
§63.1	**Applicability**
(a)	• General description of organization and relationship with other subparts under Part 63
(a)(13)	• Provisions in other relevant subparts supersede conflicting Subpart A provisions
(b)	• Initial applicability determination
(b)(2)	• Requirement to obtain Title V operating permit
(b)(3)	• Requirement to record nonaffected status
(c)	• Applicability of Subpart A for sources regulated under other Part 63 regulations
(c)(2)	• Title V permit requirements for area sources
(c)(4)	• Subpart A applicability to HAP early reduction sources
(c)(5)	• Notification requirements for area sources that become major sources
(e)	• Requirement to obtain Title V permit prior to promulgation of relevant Part 63 standards
§63.2	**Definitions**
§63.3	**Units and abbreviations**
§63.4	**Prohibited activities and circumvention**
(a)	• Requirements to operate in compliance with Part 63 regulations unless compliance extensions or exemptions have been granted; requirements to operate in compliance with Title V permit
(b)	• Prohibition of gaseous dilution to achieve compliance; prohibition of facility fragmentation to avoid regulation
§63.5	**Construction and reconstruction**
(a)(1)	• Applicability of preconstruction review to sources subject to relevant subparts of Part 63
(a)(2)	• New source construction or reconstruction of an existing source after the proposal date of a relevant standard
(b)(1)	• Affected reconstructed source becomes subject to standards for new sources
(b)(3)	• Approval by EPA or permitting authority prior to construction or reconstruction of affected source
(b)(4)	• Notification of permitting authority prior to construction or reconstruction of affected source
(b)(5)	• Prohibition of operation without complying with Part 63 requirements
(b)(6)	• Equipment additions or process changes considered part of affected source
(d)(1)	• General preconstruction application requirements; application generally required "as soon as practicable"
(d)(2)	• Technical information requirements
(d)(3)	• Additional requirements for reconstruction applications
(e)	• Approval process for construction/reconstruction applications (includes deadlines for completeness determinations, notification of intended denials, and final approval deadlines)
(f)	• Approval based on prior state preconstruction review
§63.6	**Compliance with standards and maintenance requirements**
(a)	• Applicability
(b)	• Compliance dates for new and reconstructed sources (see Table 5-9, pages 5.32–5.33)
(c)	• Compliance dates for existing sources (see Table 5-9, pages 5.32–5.33)
(e)	• Operation and maintenance requirements

TABLE 5-8
Frequently Cited Part 63, Subpart A General NESHAP Regulations—Continued

Regulatory citation	*Description*
§63.6	**Compliance with standards and maintenance requirements**—Continued
(e)(1)	• Use of good air pollution control practices
(e)(2)	• Regulatory determination of acceptable operation and maintenance based on review of procedures
(e)(3)(i)	• Development of an SSMP
(e)(3)(ii)	• Sources must operate according to an SSMP
(e)(3)(iii)	• Maintaining records to demonstrate use of SSMP procedures
(e)(3)(iv)	• Recording and reporting actions inconsistent with SSMP
(e)(3)(v)	• Sources must update SSMP and retain old SSMPs for 5 years
(e)(3)(vi)	• Standard operating procedures used in lieu of SSMP
(e)(3)(vii)	• EPA may require revisions to SSMP
(e)(3)(viii)	• Revision required within 45 days if SSMP fails to address an event that is considered a malfunction
(f)	• Compliance with nonopacity emission standards
(f)(1)	• Standards generally apply except during periods of SSM
(f)(2)	• Compliance with nonopacity standards determined based on performance tests
(g)	• Alternative nonopacity emission standard
(g)(1)	• Alternative must achieve at least an equivalent reduction in HAP emissions
(g)(2)	• Test plan submission for alternative nonopacity emission standard
(h)	• Compliance with opacity and visible emission standards
(h)(1)	• Standards generally apply except during periods of SSM
(h)(2)	• Compliance determined by 1) COMS; 2) Test Method 9 of Part 60, Appendix A (for opacity); or 3) Test Method 22 of Part 60, Appendix A (for visible emissions)
(h)(4)	• Prior notification of opacity or visible emission observations required per §63.9(f)
(h)(5)	• Requirements for conducting opacity or visible emission observations
(h)(6)	• Records of tests must be made available to EPA
(h)(7)	• COMS usage
(h)(7)(i)	• Recording and reporting COMS data
(h)(7)(ii)	• COMS data submission in lieu of Method 9 data
(h)(7)(iii)	• COMS data reduction
(h)(7)(iv)	• Demonstration that COMS data are quality assured
(h)(8)	• Finding of noncompliance
(h)(9)	• Petition for adjustment of data that fail to meet opacity emission standard
(i)	• Compliance extensions due to 1) inability to comply, 2) HAP early reductions, or 3) implementation of BACT or LAER
(j)	• Presidential exemption for national security interests
§63.7	**Performance testing requirements**
(a)	• Performance tests generally required within 180 days after becoming an affected source
(b)	• Notification of performance test required at least 60 days prior to scheduled test date
(c)(2)	• Requirement to develop and submit site-specific test plan at least 60 days prior to scheduled performance test date
(c)(3)	• Approval of site-specific test plan within 30 days after receipt of complete information

TABLE 5-8
Frequently Cited Part 63, Subpart A General NESHAP Regulations—Continued

Regulatory citation	*Description*
§63.7	**Performance testing requirements**—Continued
(c)(4)	• Performance test method audit program
(d)	• Requirement to provide adequate performance test facilities
(e)	• Performance tests conducted under representative conditions
(e)(2)	• Performance test data reduction
(e)(3)	• Requirement for three separate test runs
(f)	• Alternative test methods allowed if approved by EPA
(g)	• Data analysis, recordkeeping, and reporting
(g)(1)	• Data submission required 60 days following test completion
(g)(3)	• Record retention required for 5 years
(h)	• Performance test waiver available
§63.8	**Monitoring requirements**
(a)	• Applicability
(b)	• Monitoring to be conducted according to relevant standards
(c)	• Operation and maintenance of CMS
(c)(1)(i)	• Requirement for immediate CMS repair for routine malfunctions; report required in semiannual SSM report
(c)(1)(ii)	• Actions inconsistent with SSMP must be verbally reported within 24 hours of action initiation and followed with a report 2 weeks later detailing corrective actions
(c)(4)	• Requirement for continuous operation of CMS except for malfunctions, out-of-control periods, repairs, maintenance, and adjustments; cycling frequencies
(c)(6)	• Requirement to check zero and high-level calibration drifts
(c)(7)	• Description of CMS out-of-control periods
(c)(8)	• Requirement to report out-of-control periods
(d)	• Quality control program
(e)	• Performance evaluation of CMS
(e)(1)	• Requirement to conduct performance evaluation of CMS
(e)(2)	• Notification of performance evaluation required at least 60 days prior to scheduled test date
(e)(3)	• Submission of site-specific performance evaluation test plan
(e)(4)	• Requirement to conduct 1) CMS performance evaluation during performance test, 2) COMS performance evaluation before conducting a performance test, and 3) performance evaluation within 180 days of the compliance date if a performance test is not required
(e)(5)	• Requirement to submit performance evaluation with performance test report or within 60 days of evaluation completion if no performance test is required
(f)	• Opportunity to use alternative monitoring method
(g)	• CMS data reduction requirements
§63.9	**Notification requirements**
(a)(2)	• Exemption from §63.9 provisions for sources operating under a compliance extension
(b)(1)	• Initial notification that a source is affected applies at the time the source becomes affected under a relevant Part 63 regulation
(b)(2–3)	• Deadlines for initial notification of affected status; generally 120 days after source becomes affected

TABLE 5-8
Frequently Cited Part 63, Subpart A General NESHAP Regulations—Continued

Regulatory citation	*Description*
§63.9	**Notification requirements**—Continued
(b)(4–5)	• Notification of intent to construct or reconstruct source
(c)	• Request for compliance extension
(d)	• Notification that source is subject to special compliance requirements
(e)	• Notification of performance test required at least 60 days before scheduled test date
(f)	• Notification of opacity and visible emission observations required with notification of performance test date (or not less than 30 days prior to opacity or visible emission observations if no performance test is required)
(g)	• Notification requirements for sources with CMS required with notification of performance test (or not less than 60 days prior to CMS evaluation if no performance test is required)
(h)	• Notification of compliance status
(h)(2)	• Notification of compliance status required before a Title V permit is issued; this must be submitted within 60 days following the completion of relevant compliance demonstrations, unless a different reporting period is specified in a relevant standard
(i)	• Deadlines can be adjusted by mutual agreement between the facility and permitting authority
(j)	• Change in previously provided information must be submitted within 15 days after the change occurs
§63.10	**Recordkeeping and reporting requirements**
(a)(1)	• Recordkeeping and reporting requirements apply to all affected sources unless otherwise specified in a relevant standard
(a)(2)	• Exemption from §63.10 provisions for sources operating under a compliance extension granted per the HAP early reduction program
(a)(5)	• Opportunity to revise reporting deadlines to coincide with deadlines in an authorized state
(a)(6–7)	• Opportunity to arrange a common schedule for multiple sources subject to multiple subparts
(b)	• General recordkeeping requirements
(b)(1)	• Retain records for 5 years; 2 years onsite
(b)(2)	• Description of required records
(b)(3)	• Nonaffected sources must retain applicability determination for 5 years
(c)	• Additional recordkeeping for sources with CMS
(c)(1–15)	• Description of required information for CMS
(d)	• General reporting requirements
(d)(2)	• Reporting results of performance tests generally required within 60 days of test completion
(d)(3)	• Reporting results of opacity and visible emission observations required along with performance test results (or within 30 days of completed observations if no performance test is required)
(d)(4)	• Progress reports required from some sources, as specified in compliance extensions
(d)(5)(i)	• Periodic SSM report required on 30th day after end of calendar half for periods in which an SSM occurred
(d)(5)(ii)	• Immediate reports required when actions taken during an SSM are inconsistent with SSMP; initial report required within 2 working days after inconsistent action began, follow-up report due within 7 working days after the end of event
(e)	• Additional reporting requirements for sources with CMS
(e)(2)(i)	• CMS performance evaluation required with results of performance test
(e)(2)(ii)	• Results of COMS performance test required at least 15 days prior to performance test

TABLE 5-8
Frequently Cited Part 63, Subpart A General NESHAP Regulations—Continued

Regulatory citation	*Description*
§63.10	**Recordkeeping and reporting requirements**—Continued
(e)(3)(i)	• Excess emission and CMS performance report required semiannually unless required more frequently by a relevant standard
(e)(3)(ii)	• Request to reduce frequency of excess emission and CMS performance reports
(e)(3)(v)	• Content and submittal dates for excess emission and CMS performance reports
(e)(3)(vi)	• Summary report contents
(e)(3)(vii)	• Summary report submitted in lieu of excess emission and CMS performance reports allowed when 1) the total duration of excess emissions or parameter exceedances is less than 1% of the total operating time, and 2) CMS downtime is less than 5% of operating time
(e)(3)(viii)	• If the conditions in §63.10(e)(3)(vii) do not exist, both a summary report and excess emission and CMS performance report are required
(e)(4)	• COMS monitoring data taken during performance test must be submitted with performance test results
(f)	• Waiver of recordkeeping and reporting requirements available
§63.11	**Control device requirements**
(a)	• Control device requirements apply only to affected sources when Part 63 subpart provisions refer to this section (directly or indirectly)
(b)	• Flare requirements
(b)(1)	• Monitoring requirements
(b)(2)	• Steam-assisted, air-assisted, or nonassisted designs allowed
(b)(3)	• Operation required while emissions are flared
(b)(4)	• Operation with no visible emissions except for 5 minutes during any 2 consecutive hours
(b)(5)	• Flame presence required at all times; pilot monitoring required using thermocouple or other device
(b)(6)	• Flares can be used only when net heating value of the gas is 11.2 MJ/scm (300 Btu/scf) if flare is steam- or air-assisted; otherwise, the net heating value of the gas must be at least 7.45 MJ/scm (200 Btu/scf)
(b)(7–8)	• Constraints on exit velocity for flare systems
§63.12	**State authority and delegations**
(a)(1)	• States and other political subdivisions may require more stringent emission limitations
(a)(2)	• States may require additional permits or approvals prior to construction, reconstruction, or operation
(a)(3)	• States may require additional emission reductions under HAP early reduction program
(b)	• Delegation of authority
(c)	• Information must be submitted to authorized states as well as to EPA
§63.13	**Addresses of state air pollution control agencies and EPA regional offices**
§63.14	**Incorporation by reference**
§63.15	**Availability of information and confidentiality**

BACT = best available control technology; CMS = continuous monitoring system; COMS = continuous opacity monitoring system; HAP = hazardous air pollutant; LAER = lowest achievable emission rate; NESHAP = national emission standards for hazardous air pollutants; SSM = start-up, shutdown, or malfunction; SSMP = start-up, shutdown, and malfunction plan.

Source: Elsevier Science Inc.; adapted from 40 *CFR* Part 63, Subpart A.

TABLE 5-9
NESHAP Compliance Deadlines

Source type	Compliance date[1]
MACT-based standards for existing sources:	
• Affected sources	• Compliance date determined by EPA in the applicable Part 63 subpart, but must not be longer than 3 years after the standard's effective date[2]
• Unaffected area sources that:	
1) Become major due to increased emissions	• Compliance date set in the relevant Part 63 subpart applicable to existing area sources that become major sources[3]
2) Become major due to addition of a new affected source or by reconstruction	• The portion of the affected facility that is new or reconstructed must meet the compliance dates for new sources
MACT-based standards for new or reconstructed sources:	
• New sources that:	
1) Start operation *before* the effective date of the relevant standard (If construction or reconstruction commenced between proposal and promulgation of the relevant standard, requirement No. 3 within this category applies.)	• Effective date of the relevant standard
2) Start operation after the effective date of the relevant standard	• Start-up of the new source
3) Start construction or reconstruction between the proposal and promulgation dates of a relevant standard	• Compliance with proposed standard required upon start-up, *and* • If final regulation is more stringent than proposed regulation, compliance with final regulation required 3 years after promulgation date
• Unaffected new area source that becomes a major source	• Immediately upon becoming a major source
Health-based standards for existing sources:	
• Affected sources	• 90 days after the effective date
• Unaffected area sources that:	
1) Become major due to increased emissions	• Compliance date set in the relevant Part 63 subpart applicable to existing area sources that become major sources[3]
2) Become major due to addition of a new affected source or by reconstruction	• The portion of the affected facility that is new or reconstructed must meet the compliance dates for new sources

TABLE 5-9
NESHAP Compliance Deadlines—Continued

Source type	*Compliance date*[1]
Health-based standards for new or reconstructed sources:	
• New sources that:	
1) Start operation *before* the effective date of the relevant standard (If construction or reconstruction commenced between proposal and promulgation of the relevant standard, requirement No. 3 or 4 within this category applies.)	• Effective date of the relevant standard
2) Start operation after the effective date of the relevant standard	• Start-up of the new source
3) Facility for which construction or reconstruction started between the MACT-based standard's proposal date and the health-based standard's proposal date	• 10 years after the date construction or reconstruction began (but not before the health-based standard is promulgated)[4]
4) Facility for which construction or reconstruction started between the health-based standard's proposal date and its effective date	• Compliance with the proposed health-based standard required upon start-up, *and* • If final regulation is more stringent than proposed regulation, compliance with final regulation is required 3 years after promulgation date

MACT = maximum achievable control technology; NESHAP = national emission standards for hazardous air pollutants.

[1]Assumes the source has not received a compliance extension.

[2]Not applicable to existing coke oven battery sources, which are regulated under compliance requirements set forth in CAA Section 112(i)(8).

[3]If no such date is explicitly stated in the relevant Part 63 subpart, the compliance period will be equivalent to that allowed for other existing sources.

[4]Unless the health-based standard is promulgated more than 10 years after facility construction or reconstruction commenced; in that case, a compliance date given in either requirement No. 1 or 2 of this category applies.

Source: Elsevier Science Inc.; adapted from §63.6(b) and (c).

TABLE 5-10
Reporting Requirements Associated With the General NESHAP Provisions[1]

Reporting requirement	*Deadline*
General reporting requirements:	
• Performance test results	• 60 days after the performance test is completed
• Opacity or visible emission observation results	• To be included with performance test results submitted 60 days after the performance test is completed; if a performance test is not required, to be reported 30 days after the observations are completed
• Periodic start-up, shutdown, and malfunction reports (for actions consistent with procedures in the source's start-up, shutdown, and malfunction report)	• 30 days after the end of the calendar half[2]
• Immediate start-up, shutdown, and malfunction reports (actions that are *not* consistent with procedures in the source's start-up, shutdown, and malfunction report)	• Verbal report within 2 working days after the event begins and a written report within 7 days after the event ends
• Excess emission and monitoring system reports	• 30 days after the end of the calendar quarter or calendar half, as applicable
Additional reporting requirements for sources with CMS:	
• CMS performance evaluation results	• To be submitted along with the performance test results 60 days after the performance test is completed
• Continuous opacity monitoring system performance evaluation results (when monitoring system performance must be demonstrated prior to the source performance test)[3]	• 15 days before the performance test is conducted
• Excess emission and CMS performance reports[4]	• 30 days after the end of the calendar half[5]
• Excess emission and CMS performance summary reports	• 30 days after the end of the calendar half[5]

CMS = continuous monitoring system; NESHAP = national emission standards for hazardous air pollutants.

[1]Notification requirements are not included.

[2]Unless more frequent submissions are specified in a relevant Part 63 standard or otherwise established in the source's Title V permit. Note that this report is waived if start-ups, shutdowns, or malfunctions occurred during the reporting period.

[3]Applicable to affected sources using a continuous opacity monitoring system to determine opacity compliance during a performance test.

[4]Not necessary if 1) the total duration of excess emissions or process/control system parameter exceedances for the reporting period is less than 1% of the total operating time for the reporting period, and 2) CMS downtime for the reporting period is less than 5% of the total operating time for the reporting period. Instead, only a summary report is necessary.

[5]Unless 1) the CMS data are to be used directly for compliance determination and the source had excess emissions, in which case quarterly reports will be necessary; 2) more frequent submissions are specified in a relevant Part 63 standard; or 3) EPA determines that more frequent reporting is necessary.

Source: Elsevier Science Inc.; adapted from §63.10(d) and (e).

TABLE 5-11
General NESHAP Performance Test Schedules

Affected source type	*Performance test deadline*
Existing source	
—That begins operation on or *before* the effective date of a relevant standard	• 180 days after the compliance date established in the specific source-category regulation[1]
—That begins operation *after* the effective date of a relevant standard	• 180 days after initial start-up
New source	
—With an initial start-up *before* the effective date of the relevant standard	• 180 days after the effective date
—With an initial start-up *after* the effective date of the relevant standard	• 180 days after initial start-up
—For which construction or reconstruction begins after a MACT-based standard is proposed and before a health-based standard is proposed[2]	• 180 days after the compliance date
Source that obtains a compliance extension	• 180 days after the compliance extension expires
Requirements for certain new or reconstructed sources if the final source-category standard is more stringent than the proposed standard:	
—New or reconstructed source for which 1) construction or reconstruction began between the standard's proposal and promulgation dates, and 2) the source's start-up date occurs *before* the effective date of the standard	• Performance tests (as required by the proposed standard) must be completed 180 days after the effective date of the promulgated standard, *and* • Performance tests (as required by the final standard) must be completed 3 years and 180 days after the effective date[3]
—New or reconstructed source for which 1) construction or reconstruction began between the standard's proposal and promulgation dates, and 2) the source's start-up date occurs *after* the effective date of the standard	• Performance tests (as required by the proposed standard) must be completed 180 days after the source's start-up date, *and* • Performance tests (as required by the final standard) must be completed 3 years and 180 days after start-up[3]

MACT = maximum achievable control technology; NESHAP = national emission standards for hazardous air pollutants.

[1]For both MACT (technology-based) standards and health-based standards.

[2]For sources subject to health-based standards only.

[3]Alternatively, a source owner/operator can conduct the first performance test to demonstrate compliance with the *final* regulation; in this case, the second performance test is not required.

Source: Elsevier Science Inc.; adapted from §63.7(a)(2).

TABLE 5-12
Toxic Substances Regulated Under the Accidental Release Prevention Program

Chemical	*CAS no.*	*TQ (lbs)*	*Toxic endpoint (mg/L)*
Acrolein	107-02-8	5,000	0.0011
Acrylonitrile	107-13-1	20,000	0.076
Acrylyl chloride	814-68-6	5,000	0.00090
Allyl alcohol	107-18-6	15,000	0.036
Allylamine	107-11-9	10,000	0.0032
Ammonia (anhydrous)	7664-41-7	10,000	0.14
Ammonia (aqueous solution, conc. ≥20%)	7664-41-7	20,000	0.14
Arsenous trichloride	7784-34-1	15,000	0.010
Arsine	7784-42-1	1,000	0.0019
Boron trichloride	10294-34-5	5,000	0.010
Boron trifluoride	7637-07-2	5,000	0.028
Boron trifluoride compound with methyl ether (1:1)	353-42-4	15,000	0.023
Bromine	7726-95-6	10,000	0.0065
Carbon disulfide	75-15-0	20,000	0.16
Chlorine	7782-50-5	2,500	0.0087
Chlorine dioxide	10049-04-4	1,000	0.0028
Chloroform	67-66-3	20,000	0.49
Chloromethyl ether	542-88-1	1,000	0.00025
Chloromethyl methyl ether	107-30-2	5,000	0.0018
Crotonaldehyde	4170-30-3	20,000	0.029
Crotonaldehyde, (E)-	123-73-9	20,000	0.029
Cyanogen chloride	506-77-4	10,000	0.030
Cyclohexylamine	108-91-8	15,000	0.16
Diborane	19287-45-7	2,500	0.0011
Dimethyldichlorosilane	75-78-5	5,000	0.026
1,1-Dimethylhydrazine	57-14-7	15,000	0.012
Epichlorohydrin	106-89-8	20,000	0.076
Ethylene oxide	75-21-8	10,000	0.090
Ethylenediamine	107-15-3	20,000	0.49
Ethyleneimine	151-56-4	10,000	0.018
Fluorine	7782-41-4	1,000	0.0039
Formaldehyde	50-00-0	15,000	0.012
Furan	110-00-9	5,000	0.0012
Hydrazine	302-01-2	15,000	0.011
Hydrochloric acid (solution, conc. ≥37%)	7647-01-0	15,000	0.030
Hydrocyanic acid	74-90-8	2,500	0.011
Hydrogen chloride (anhydrous)	7647-01-0	5,000	0.030
Hydrogen fluoride/hydrofluoric acid (conc. ≥50%)	7664-39-3	1,000	0.016
Hydrogen selenide	7783-07-5	500	0.00066
Hydrogen sulfide	7783-06-4	10,000	0.042

TABLE 5-12
Toxic Substances Regulated Under the Accidental Release Prevention Program—Continued

Chemical	*CAS no.*	*TQ (lbs)*	*Toxic endpoint (mg/L)*
Iron, pentacarbonyl-	13463-40-6	2,500	0.00044
Isobutyronitrile	78-82-0	20,000	0.14
Isopropyl chloroformate	108-23-6	15,000	0.10
Methacrylonitrile	126-98-7	10,000	0.0027
Methyl chloride	74-87-3	10,000	0.82
Methyl chloroformate	79-22-1	5,000	0.0019
Methyl hydrazine	60-34-4	15,000	0.0094
Methyl isocyanate	624-83-9	10,000	0.0012
Methyl mercaptan	74-93-1	10,000	0.049
Methyl thiocyanate	556-64-9	20,000	0.085
Methyltrichlorosilane	75-79-6	5,000	0.018
Nickel carbonyl	13463-39-3	1,000	0.00067
Nitric acid (conc. ≥80%)	7697-37-2	15,000	0.026
Nitric oxide	10102-43-9	10,000	0.031
Oleum (fuming sulfuric acid)	8014-95-7	10,000	0.010
Peracetic acid	79-21-0	10,000	0.0045
Perchloromethylmercaptan	594-42-3	10,000	0.0076
Phosgene	75-44-5	500	0.00081
Phosphine	7803-51-2	5,000	0.0035
Phosphorus oxychloride	10025-87-3	5,000	0.0030
Phosphorus trichloride	7719-12-2	15,000	0.028
Piperidine	110-89-4	15,000	0.022
Propionitrile	107-12-0	10,000	0.0037
Propyl chloroformate	109-61-5	15,000	0.010
Propylene oxide	75-56-9	10,000	0.59
Propyleneimine	75-55-8	10,000	0.12
Sulfur dioxide (anhydrous)	7446-09-5	5,000	0.0078
Sulfur tetrafluoride	7783-60-0	2,500	0.0092
Sulfur trioxide	7446-11-9	10,000	0.010
Tetramethyllead	75-74-1	10,000	0.0040
Tetranitromethane	509-14-8	10,000	0.0040
Titanium tetrachloride	7550-45-0	2,500	0.020
Toluene 2,4-diisocyanate	584-84-9	10,000	0.0070
Toluene 2,6-diisocyanate	91-08-7	10,000	0.0070
Toluene diisocyanate (unspecified isomer)	26471-62-5	10,000	0.0070
Trimethylchlorosilane	75-77-4	10,000	0.050
Vinyl acetate monomer	108-05-4	15,000	0.26

CAS = Chemical Abstract Service; TQ = threshold quantity.

Source: 40 *CFR* §68.130, Table 1 and Appendix A to 40 *CFR* Part 68.

TABLE 5-13
Flammable Substances Regulated Under the Accidental Release Prevention Program[1,2]

Chemical	CAS no.	Chemical	CAS no.
Acetaldehyde	75-07-0	Isopropyl chloride	75-29-6
Acetylene	74-86-2	Isopropylamine	75-31-0
Bromotrifluorethylene	598-73-2	Methane	74-82-8
1,3-Butadiene	106-99-0	3-Methyl-1-butene	563-45-1
Butane	106-97-8	2-Methyl-1-butene	563-46-2
Butene	25167-67-3	Methyl ether	115-10-6
1-Butene	106-98-9	Methyl formate	107-31-3
2-Butene	107-01-7	Methylamine	74-89-5
2-Butene-*cis*	590-18-1	2-Methylpropene	115-11-7
2-Butene-*trans*	624-64-6	1,3-Pentadiene	504-60-9
Carbon oxysulfide	463-58-1	Pentane	109-66-0
Chlorine monoxide	7791-21-1	1-Pentene	109-67-1
2-Chloropropylene	557-98-2	2-Pentene, (E)-	646-04-8
1-Chloropropylene	590-21-6	2-Pentene, (Z)-	627-20-3
Cyanogen	460-19-5	Propadiene	463-49-0
Cyclopropane	75-19-4	Propane	74-98-6
Dichlorosilane	4109-96-0	Propylene	115-07-1
Difluoroethane	75-37-6	Propyne	74-99-7
Dimethylamine	124-40-3	Silane	7803-62-5
2,2-Dimethylpropane	463-82-1	Tetrafluoroethylene	116-14-3
Ethane	74-84-0	Tetramethylsilane	75-76-3
Ethyl acetylene	107-00-6	Trichlorosilane	10025-78-2
Ethyl chloride	75-00-3	Trifluorochloroethylene	79-38-9
Ethyl ether	60-29-7	Trimethylamine	75-50-3
Ethyl mercaptan	75-08-1	Vinyl acetylene	689-97-4
Ethyl nitrite	109-95-5	Vinyl chloride	75-01-4
Ethylamine	75-04-7	Vinyl ethyl ether	109-92-2
Ethylene	74-85-1	Vinyl fluoride	75-02-5
Hydrogen	1333-74-0	Vinyl methyl ether	107-25-5
Isobutane	75-28-5	Vinylidene chloride	75-35-4
Isopentane	78-78-4	Vinylidene fluoride	75-38-7
Isoprene	78-79-5		

CAS = Chemical Abstract Service.

[1]The threshold quantity for all regulated flammable substances is 10,000 pounds.

[2]The endpoints for flammables vary depending on the hazard under review. For the explosion hazard, the endpoint is an overpressure of 1 psi. For explosion/fire potential, the endpoint is the lower explosive limit for the substance. For radiant heat hazard, the endpoint is exposure to radiant heat of 5 kw/m^2 for 40 seconds.

Source: 40 *CFR* §§68.22(a) and 68.130, Table 3.

TABLE 5-14

Summary of Accidental Release Prevention Program Levels

Program level	*Criteria*	*RMP requirements*
Program 3—Significant risk	• Processes are classified as Program 3 processes if 1) they cannot qualify for Program 1, *and* 2) they belong to one of nine specific SIC codes or they are subject to the OSHA process safety management standard (29 *CFR* §1910.119). The nine SIC codes are 2611 (pulp mills), 2812 (chlor-alkali processes), 2819 (industrial inorganic processes), 2821 (plastics and resins processes), 2865 (cyclic crude intermediates processes), 2869 (industrial organics processes), 2873 (nitrogen fertilizer processes), 2879 (agricultural chemicals), and 2911 (refineries).	Full RMP
Program 2—Intermediate risk	• Any process that does not qualify for Program 1 designation and is not required to be classified as Program 3 is considered a Program 2 process.	Streamlined RMP
Program 1—Minimal risk	• In the 5 years prior to submission of the RMP, the process must not have experienced an accidental release of a regulated substance "where exposure to the substance, its reaction products, overpressure generated by an explosion involving the substance, or radiant heat generated by a fire involving the substance [§68.10(b)(1)]" led to death, injury, or response/restoration activities for exposure of an environmental receptor • The distance to a public receptor (e.g., residence, school, office, or park) must be greater than the distance to the toxic or flammable substance endpoint that results from the worst-case release scenario. Endpoints for toxics are compound-specific and are listed in Table 5-12 (pages 5.36 and 5.37). The endpoints for flammables vary depending on the hazard under study and are summarized in Footnote 2 of Table 5-13 (page 5.38). • Emergency response procedures must have been coordinated between the facility and local emergency planning and response organizations.	Minimal RMP

OSHA = Occupational Safety and Health Administration; RMP = risk management plan; SIC = Standard Industrial Classification.

Source: Elsevier Science Inc.

FIGURE 5-1

Determining the Level of Regulation Under the Accidental Release Prevention Program

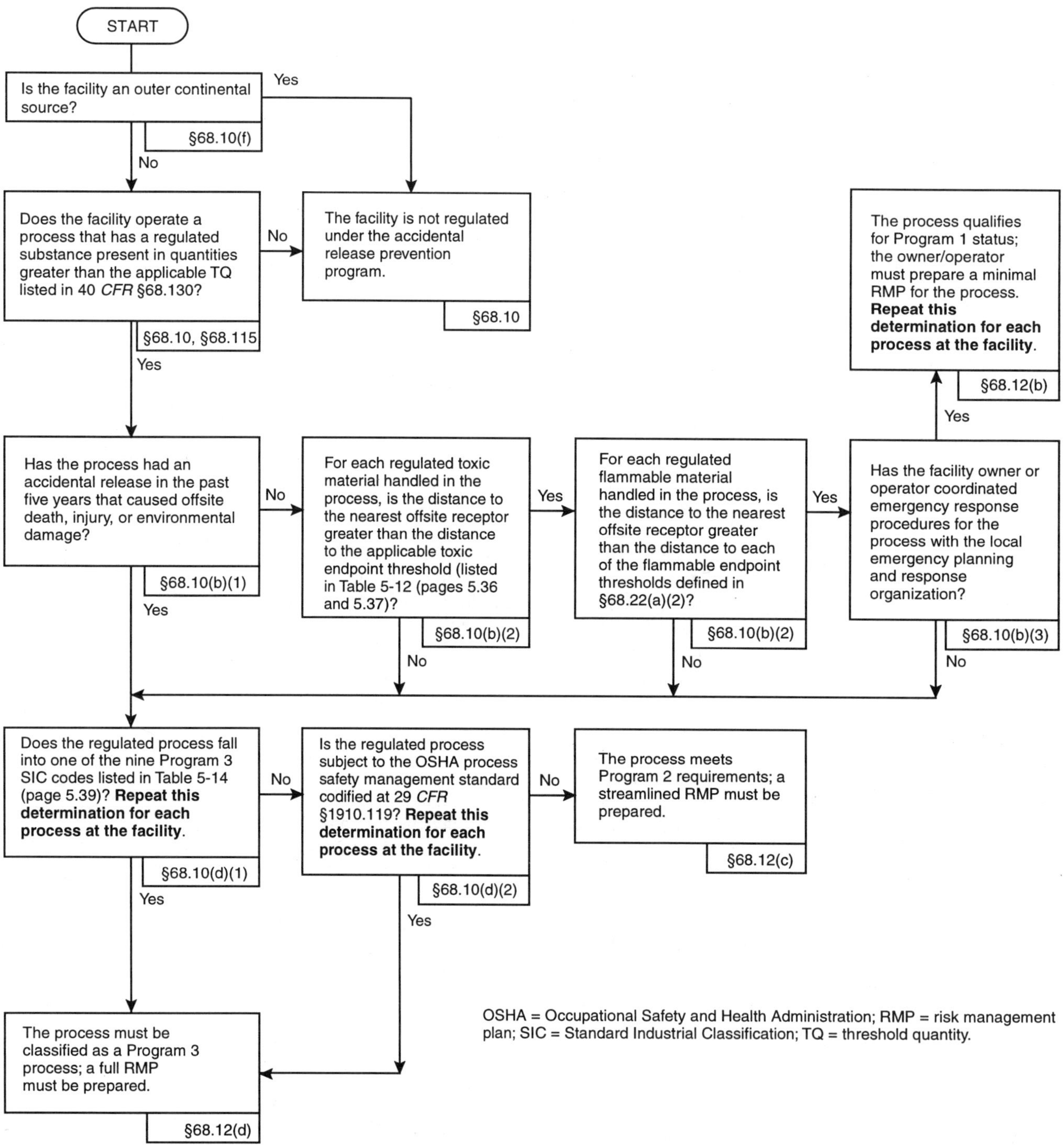

OSHA = Occupational Safety and Health Administration; RMP = risk management plan; SIC = Standard Industrial Classification; TQ = threshold quantity.

Source: Elsevier Science Inc.

6 The Operating Permit Program

Title V operating permits, as mandated by the 1990 Amendments, serve as "umbrella permits" that incorporate all applicable CAA requirements for each regulated source. As shown in Figure 6-1, enforceable conditions included in a given facility's operating permit may reference any and all requirements promulgated under the CAA.

Congress stipulated that states and authorized local agencies implement the Title V operating permit program based on minimum federal requirements issued by EPA. The agency promulgated such regulations on July 21, 1992 (57 *FR* 32250) as Part 70 to Title 40 of the *CFR*. Key elements of the Part 70 regulations are listed in Table 6-1 (page 6.2). (Note that the Part 70 regulations have been at the center of controversy and a number of lawsuits since they were issued. EPA has put forth several revision proposals to try to strike a balance between competing interests. Although numerous issues are involved, the primary controversy centers on balancing a source's operational flexibility with the public's opportunity for comment. See "Additional References" on page 6.13 for more information on this subject.)

In cases where a given state or local permitting authority is overly tardy in preparing its Title V program, or where such programs contain terms or conditions that EPA views as unacceptable, EPA must step in and implement a federal Title V program. Regulations defining such federal Title V program rules were promulgated as 40 *CFR* Part 71 on July 1, 1996 (61 *FR* 34202).

FIGURE 6-1

Applicable Requirements Incorporated Into Title V Operating Permits

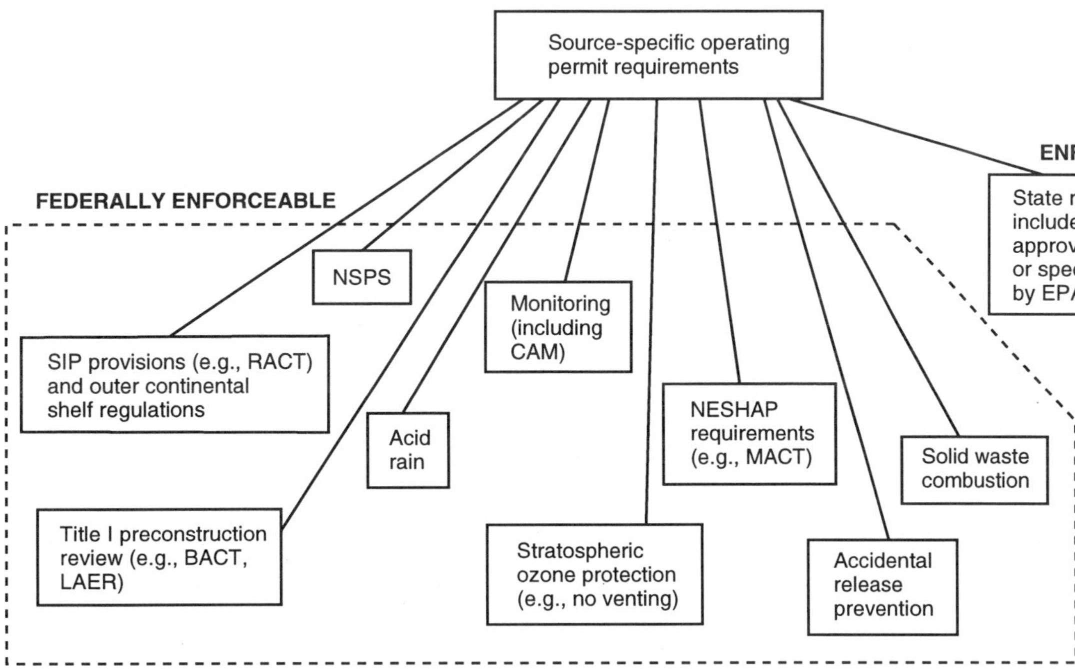

CAM = compliance assurance monitoring; BACT = best available control technology; LAER = lowest achievable emission rate; MACT = maximum achievable control technology; NESHAP = national emission standards for hazardous air pollutants; NSPS = new source performance standards; RACT = reasonably available control technology; SIP = state implementation plan.

Source: Elsevier Science Inc.

TABLE 6-1
Key Operating Permit Program *CFR* References

Topic	*40 CFR references*
Definitions	§70.2
Applicability	
Stationary sources subject to Part 70 rules	§70.3(a)
Source category exemptions	§70.3(b)
Permit applications	
Deadlines	§70.5(a)(1)
Application forms and required content	§70.5(c)
Compliance plans and schedules	§70.5(c)(8)
Compliance certification	§70.5(c)(9)
Permit terms and conditions	
Monitoring and related record-keeping and reporting	§70.6(a)(3)
Emission trading	§70.6(a)(8)
Operating flexibility	§70.6(a)(9)
General permits	§70.6(d)
Temporary sources	§70.6(e)
Permit shields	§70.6(f)
Permit revisions	
Administrative amendments	§70.7(d)
Minor modifications	§70.7(e)(2)
Group processing of minor revisions	§70.7(e)(3)
Significant modifications	§70.7(e)(4)

Source: Elsevier Science Inc.

TABLE 6-2
Sources Subject to the Operating Permit Program

Any "major" source, as defined below:
- Any stationary source (or group of sources located within a contiguous area and under common control) that has the potential to emit 100 tpy or more of any air pollutant,
- A "major" stationary source as defined under Title I nonattainment NSR or PSD provisions, or
- A "major" HAP stationary source having the potential to emit 10 tpy or more of any single HAP, or 25 tpy or more of total HAPs

Any source (including area sources) regulated by NSPS[1,2]

Any source (including area sources) subject to the NESHAP program except those sources 1) subject only to accidental release prevention provisions, and 2) in source categories where EPA has waived Title V applicability[1,3]

Any affected source under the acid rain program[4]

Any source in a source category otherwise designated by EPA

HAP = hazardous air pollutant; NESHAP = national emission standards for hazardous air pollutants; NSPS = new source performance standards; NSR = new source review; PSD = prevention of significant deterioration; tpy = tons per year.

[1]In most cases, area sources need not apply for an operating permit until five years after the applicable operating permit program has been approved.

[2]If the source is subject only to the NSPS for residential wood heaters, it is exempt from the operating permit program.

[3]If the source is subject only to the asbestos demolition and renovation NESHAP, it is exempt from the operating permit program.

[4]Special permit application deadlines apply to affected units under the acid rain program.

Source: Elsevier Science Inc.

Applicability

Table 6-2 contains a list of the types of sources subject to the operating permit program; in turn, Figures 6-2(a) and 6-2(b) (pages 6.3 and 6.4, respectively) illustrate a process for determining whether a source or group of sources must obtain such a permit. [Note that Table 6-3 (page 6.5) contains a list of source categories for which fugitive emissions must be considered in major source determinations; this table is referenced in Figure 6-2(b).]

Generally, sources subject to Title V permitting requirements must submit a permit application within one year of becoming subject to a Part 70 program. Specific elements that must be included in a source's Part 70 operating permit application are listed in Table 6-4 (page 6.5).

EPA has issued two guidance documents to reduce the complexity and preparation costs associated with Title V operating permits. The guidance documents, known as the "White Paper" and "White Paper Number 2" were issued in July 1995 and March 1996, respectively. *APC* articles describing these important guidance documents are provided under "Additional References" on page 6.13.

General Permits

Because such a large number of facilities will eventually be required to obtain operating permits (about 68,000 sources),

FIGURE 6-2(a)
Determining Applicability of the Operating Permit Program

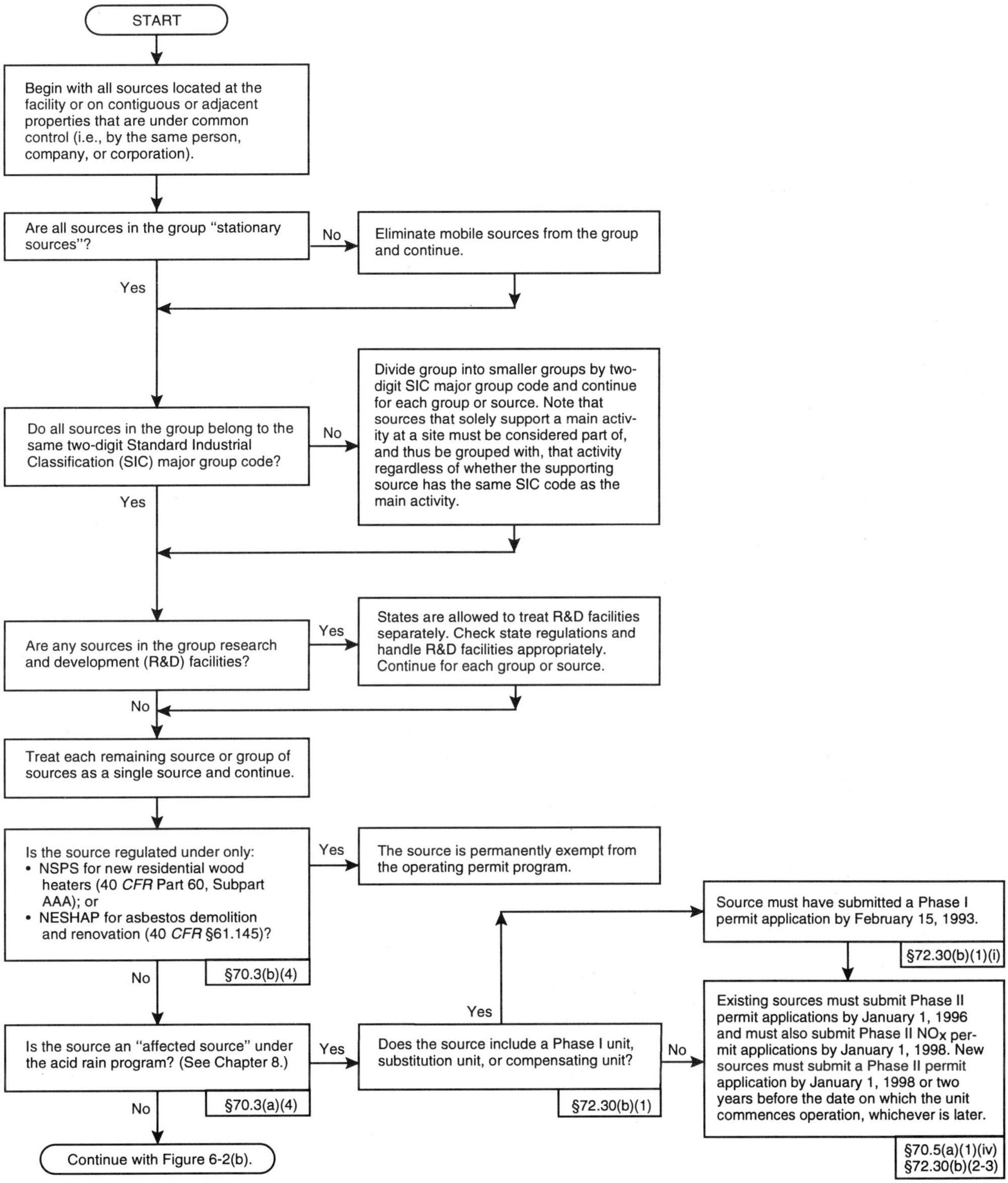

Source: Elsevier Science Inc.

FIGURE 6-2(b)
Determining Applicability of the Operating Permit Program

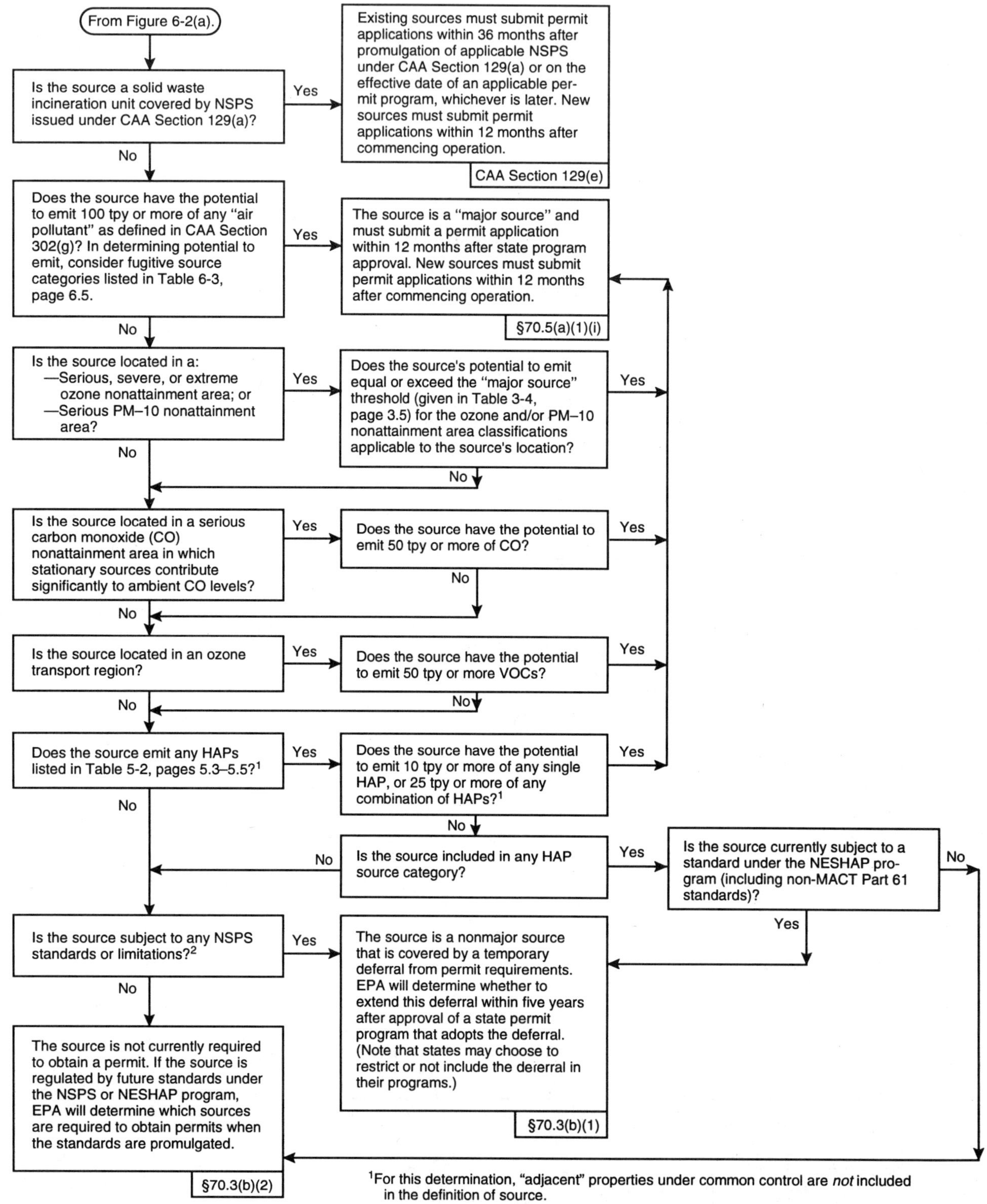

[1] For this determination, "adjacent" properties under common control are *not* included in the definition of source.

[2] Some NSPS exempt certain sources from Title V permitting; for example, see 40 *CFR* Part 60, Subpart WWW [§60.752(a)] for municipal solid waste landfills.

Source: Elsevier Science Inc.

TABLE 6-3
Source Categories for Which Fugitive Emissions Must Be Considered in Major Source Determinations

Carbon black plants (furnace process)

Charcoal production plants

Chemical process plants

Coal cleaning plants (with thermal dryers)

Coke oven batteries

Fossil fuel boilers (or combinations thereof) totaling more than 250 MMBtu/hr heat input

Fossil fuel-fired steam electric plants of more than 250 MMBtu/hr heat input

Fuel conversion plants

Glass fiber processing plants

Hydrofluoric, sulfuric, or nitric acid plants

Iron and steel mills

Kraft pulp mills

Lime plants

Municipal incinerators capable of charging more than 250 tons of refuse per day

Petroleum refineries

Petroleum storage and transfer units with a total storage capacity exceeding 300,000 barrels

Phosphate rock processing plants

Portland cement plants

Primary aluminum ore reduction plants

Primary copper smelters

Primary lead smelters

Primary zinc smelters

Secondary metal production plants

Sintering plants

Sulfur recovery plants

Taconite ore processing plants

Any other stationary source category regulated by a standard promulgated under CAA Section 111 or 112 as of August 7, 1980.[1]

NESHAP = national emission standards for hazardous air pollutants; NSPS = new source performance standards.

[1]CAA Sections 111 and 112 establish the NSPS and NESHAP programs, respectively.

Source: 40 *CFR* §70.2, "Major source."

TABLE 6-4
Operating Permit Application Requirements

- General company information, including facility name and address, owner's name and agent, and facility contact person and telephone number
- Description of the source's processes and products by SIC code, including those associated with alternate scenarios
- Description of all emission points
- Emission inventory for all regulated air pollutants, including fugitive emissions from all emission units, and a listing of all pollutants for which the facility is classified as a major source
- Emission rates in tpy
- Description of fuels and how they are used, as well as a description of raw materials used, to the extent needed to determine or to regulate emissions
- Description of all air pollution control and compliance monitoring devices or activities, along with brief descriptions of operation, maintenance, and quality assurance procedures
- Description of production rates, operating schedules, and any operating limitations or restrictions on work practices that affect emissions of regulated air pollutants
- List of all applicable state and federal air pollution control requirements, including those promulgated at the time of application that will become effective during the term of the permit
- Compliance plan for all sources within the facility regulated under the permit program
- Certification of compliance with all applicable requirements and a schedule for submission of at least annual compliance certifications during the permit term
- Description of methods used to determine compliance, including monitoring, recordkeeping, reporting, and testing methods
- Information on alternative operating scenarios defined under "operational flexibility" provisions
- Supporting calculations for submitted information

SIC = Standard Industrial Classification; tpy = tons per year.

Source: Elsevier Science Inc.; compiled from 40 *CFR* §70.5(c).

EPA will allow permitting authorities to issue "general" permits covering numerous similar sources. For example, general permits could be issued for gas stations, dry cleaners, or small printers. According to EPA, the primary purpose of this approach is to provide an alternative means for permitting sources for which the normal permitting process would be overly burdensome, such as area (i.e., nonmajor) sources under the NESHAP program. To assist states and localities in implementing Title V, EPA is developing model general permits for certain source categories; these are listed in Table 6-5.

Permit Revisions

Current operating permit regulations (both the Parts 70 and 71 versions) include three means of revising operating permits: 1) administrative amendments, 2) minor permit revisions, and 3) significant permit revisions. Determining which permit revision requirement, if any, applies to a proposed facility change is one of the most complex tasks imposed by the operating permit program. Figures 6-3(a) and 6-3(b) (pages 6.7 and 6.8, respectively) present a logic diagram for

TABLE 6-5

Source Categories for Which EPA Has Issued Draft Model Permits[1]

Draft NESHAP model permits
- Inorganic arsenic from glass manufacturing plants
- Sources emitting:
- —Ethylene dichloride (from PVC plants)
- —Mercury
- —Vinyl chloride (from PVC plants)

Draft PM-10 model permits
- Calciners and dryers
- Elemental phosphorus plants
- Glass manufacturing[2]
- Grain milling
- Gypsum processing and products
- Hot mix asphalt plants
- Iron and steel mills[2,3]
- Lime kilns
- Marine grain elevators
- Nonferrous smelters[2]
- Nonmetallic minerals processing
- Nonprocess fugitive emission sources
- Paint manufacturing[2]
- Petroleum refineries
- Portland cement plants[2]
- Primary aluminum plants
- Pulp mills
- Utility boilers

Draft lead model permits
- Lead acid battery manufacturing
- Secondary lead remelt and consuming plants
- Secondary lead smelting and refining plants

Draft VOC RACT model permits
- Auto and light-duty truck coating operations
- Bulk gasoline plants and terminals
- Coating operations (can, coil, paper, fabric, flat wood paneling, vinyl, metal furniture, large appliances, magnetic wire, and miscellaneous metal parts)
- Cutback and emulsified asphalt
- Gasoline dispensing (Stage I vapor recovery)
- Graphic arts
- High-density polyethylene, polypropylene, and polystyrene resins manufacturing
- Leaking sources (petroleum refinery, natural gas/gasoline processing, synthetic organic chemical manufacturing, and polymer and resin manufacturing)
- Leaks from gasoline tank trucks
- Petroleum refineries
- Petroleum solvent and/or perchloroethylene dry cleaners
- Petroleum storage in external floating roof tanks and/or fixed-roof tanks
- Pneumatic rubber tire manufacturing
- Solvent metal cleaning (degreasing)
- Synthesized pharmaceutical products manufacturing
- Synthetic organic chemical manufacturing industry—air oxidation

Draft SO_2 model permit
- Utility boilers

NESHAP = national emission standards for hazardous air pollutants; PM-10 = particulate matter with an aerodynamic diameter less than or equal to 10 microns; PVC = polyvinyl chloride; RACT = reasonably available control technology; SO_2 = sulfur dioxide; VOC = volatile organic compound.

[1]Draft models are available from the Office of Air Quality Planning & Standards TTNWeb, OAR P&G Web site (http://www.epa.gov/ttn/caaa/amend.html) under Title V Policy and Guidance Documents.

[2]Lead included.

[3]Sources with new source performance standards.

Source: Elsevier Science Inc.

FIGURE 6-3(a)

Determining Operating Permit Revision Requirements

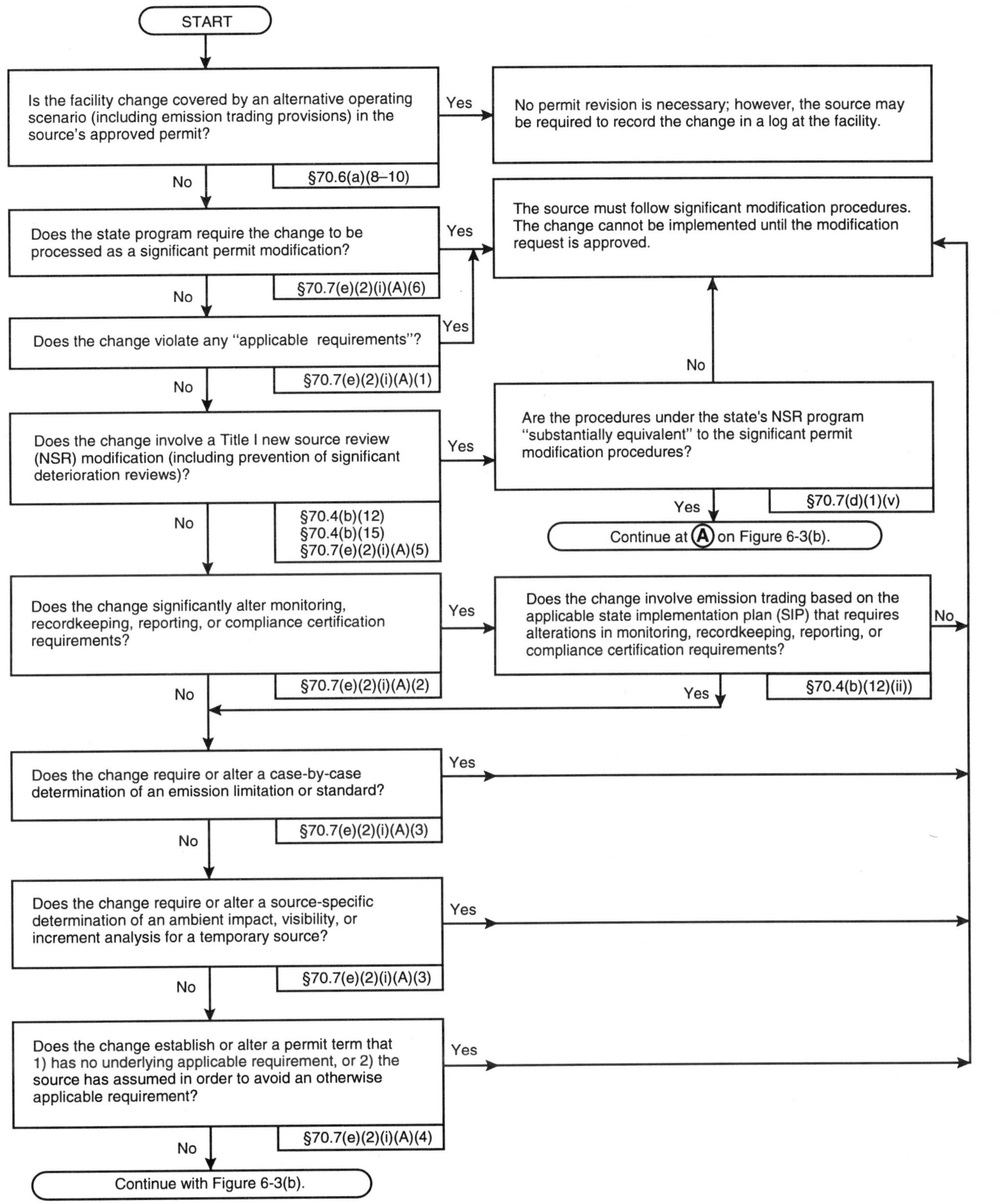

Source: Elsevier Science Inc.

FIGURE 6-3(b)

Determining Operating Permit Revision Requirements

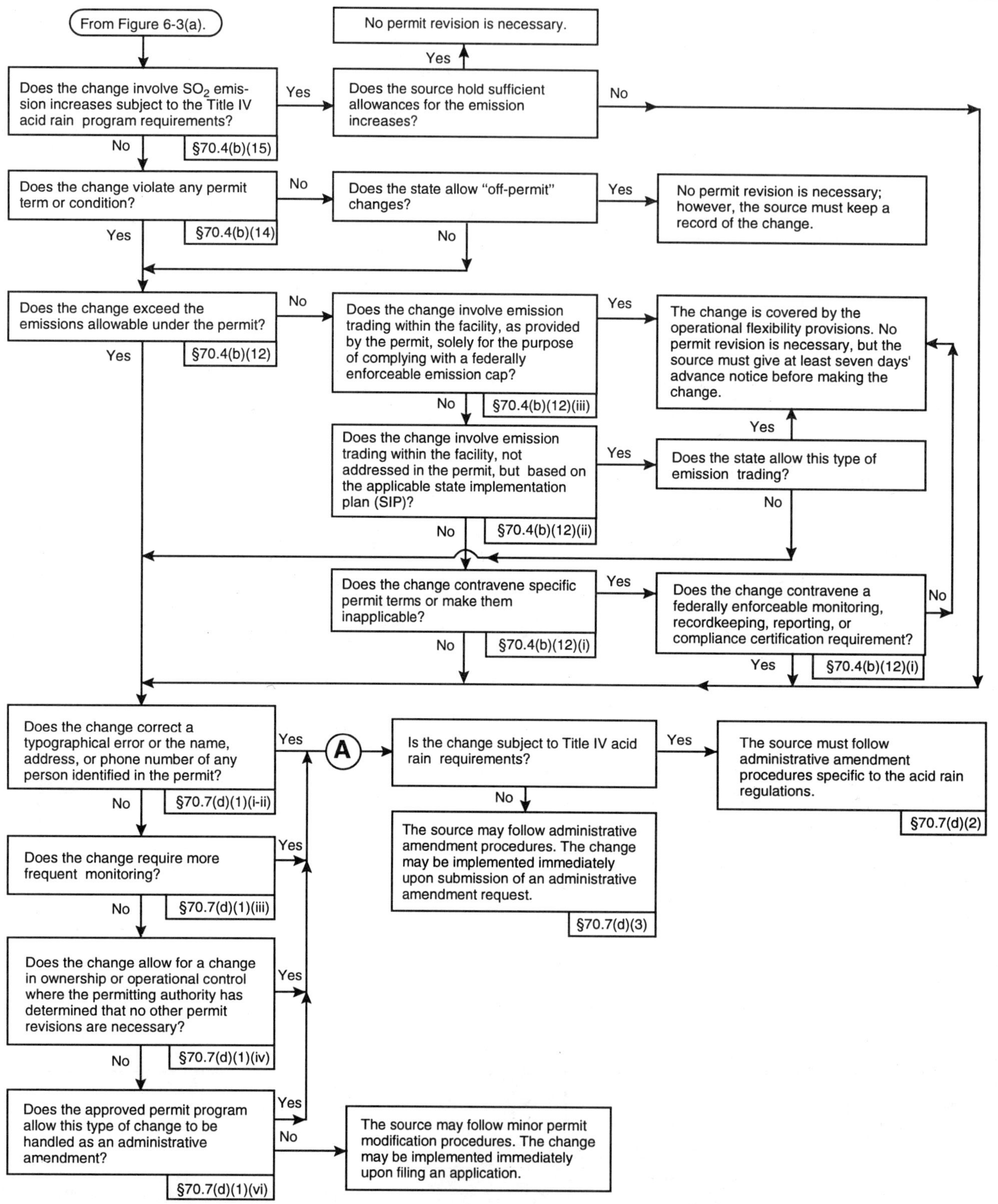

Source: Elsevier Science Inc.

determining the minimum permit revision requirements for a planned facility change. Note, however, that a given state or local program may require that a source follow a more stringent revision track than called for under the federal regulations.

Status of State and Local Title V Programs

As noted above, Congress mandated that Title V permit programs be implemented by state or local agencies. Toward that end, the 1990 amendments called for all states (and authorized local agencies) to submit operating permit programs for EPA approval by November 15, 1993. EPA was then to examine such programs to ensure that they were at least as stringent as called for in Part 70 regulations. Based on this analysis, the federal agency was then to issue one of three rulings on each program:

1. *Full approval* to those programs that met Part 70 requirements;
2. *Interim approval* to those programs with deficiencies that required correction, but that "substantially" met the requirements of Part 70; and
3. *Disapproval* to those programs that did not meet Part 70 requirements.

Although nearly every aspect of the Title V program submission and approval process is well behind schedule, all areas of the country have now obtained final approval of their Title V programs. In the majority of cases, and as shown in Table 6-6, most of those actions are *interim* approvals. Based on the provisions of CAA Section 502(g), EPA has been granting states two years to move from interim to full program approval. However, in light of continuing difficulties in revising Title V program rules, interim approval expiration deadlines have been extended several times. The latest action, taken on July 27, 1998 (63 *FR* 40054), extended the interim approval expiration deadline for *all* interim approved programs until June 1, 2000. (EPA's main motivation in issuing these extensions is to allow states to revise their programs one single time to address both the shortcomings that prompted interim approval *and* the changes that will be required when Part 70 revisions are finally issued.)

TABLE 6-6
Status of State and Local Operating Permit Programs

State/local agency	*Final action*	*FR reference*
Alabama	Interim approval[1] 11/15/95	(60 *FR* 57346)
City of Huntsville	Interim approval 11/15/95	(60 *FR* 57346)
Jefferson County	Interim approval[1] 11/15/95	(60 *FR* 57346)
Alaska	Interim approval 12/05/96	(61 *FR* 64463)
American Samoa	Conditional exemption 11/13/96	(61 *FR* 58284)
Arizona	Interim approval 10/30/96	(61 *FR* 55910)
Maricopa County	Same as above	Same as above
Pima County	Same as above	Same as above
Pinal-Gila Counties	Same as above	Same as above
Arkansas	Interim approval 09/08/95	(60 *FR* 46771)
California		
Amador County	Interim approval[1] 05/03/95	(60 *FR* 21720)
Bay Area	Interim approval[1] 06/23/95	(60 *FR* 32606)
Butte County	Interim approval[1] 05/03/95	(60 *FR* 21720)
Calaveras County	Interim approval[1] 05/03/95	(60 *FR* 21720)
Colusa County	Interim approval[1] 05/03/95	(60 *FR* 21720)
El Dorado County	Interim approval[1] 05/03/95	(60 *FR* 21720)
Feather River County	Interim approval[1] 05/03/95	(60 *FR* 21720)

TABLE 6-6
Status of State and Local Operating Permit Programs—Continued

State/local agency	Final action	FR reference
California—Continued		
Glenn County	Interim approval[1] 07/13/95	(60 *FR* 36065)
Great Basin Unified	Interim approval[1] 05/03/95	(60 *FR* 21720)
Imperial County	Interim approval[1] 05/03/95	(60 *FR* 21720)
Kern County	Interim approval[1] 05/03/95	(60 *FR* 21720)
Lake County	Interim approval[1] 07/13/95	(60 *FR* 36065)
Lassen County	Interim approval[1] 05/03/95	(60 *FR* 21720)
Mariposa County	Interim approval[1] 12/07/95	(60 *FR* 62758)
Mendocino County	Interim approval[1] 05/03/95	(60 *FR* 21720)
Modoc County	Interim approval[1] 05/03/95	(60 *FR* 21720)
Mojave Desert	Interim approval[1] 02/05/96	(61 *FR* 4217)
Monterey Bay Unified	Interim approval[1] 10/06/95	(60 *FR* 52332)
North Coast Unified	Interim approval[1] 05/03/95	(60 *FR* 21720)
Northern Sierra	Interim approval[1] 05/03/95	(60 *FR* 21720)
Northern Sonoma County	Interim approval[1] 05/03/95	(60 *FR* 21720)
Placer County	Interim approval[1] 05/03/95	(60 *FR* 21720)
Sacramento Metropolitan	Interim approval[1] 08/04/95	(60 *FR* 39862)
San Diego County	Interim approval[1] 12/07/95	(60 *FR* 62753)
San Joaquin Unified	Interim approval[1] 04/24/96	(61 *FR* 18083)
San Luis Obispo County	Interim approval[1] 11/01/95	(60 *FR* 55460)
Santa Barbara County	Interim approval[1] 11/01/95	(60 *FR* 55460)
Shasta County	Interim approval[1] 07/13/95	(60 *FR* 36065)
Siskiyou County		
South Coast	Interim approval[1] 02/27/97	(62 *FR* 8878)
Tehama County	Interim approval[1] 07/13/95	(60 *FR* 36065)
Tuolumne County	Interim approval[1] 05/03/95	(60 *FR* 21720)
Ventura County	Interim approval[1] 11/01/95	(60 *FR* 55460)
Yolo-Solano	Interim approval[1] 05/03/95	(60 *FR* 21720)
Colorado	Interim approval 01/24/95	(60 *FR* 4563)
Connecticut	Interim approval 03/24/97	(62 *FR* 13830)
Delaware	Interim approval 12/04/95	(60 *FR* 62032)
District of Columbia	Interim approval 08/07/95	(60 *FR* 40101)
Florida	Interim approval 09/25/95	(60 *FR* 49343)
Georgia	Interim approval 11/22/95	(60 *FR* 57836)
Guam	Conditional extension 11/13/96	(61 *FR* 58284)
Hawaii	Interim approval 12/01/94	(59 *FR* 61549)
Idaho	Interim approval 12/06/96	(61 *FR* 64622)
Illinois	Interim approval 03/07/95	(60 *FR* 12478)
Indiana	Interim approval 11/14/95	(60 *FR* 57188)

TABLE 6-6
Status of State and Local Operating Permit Programs—Continued

State/local agency	*Final action*	*FR reference*
Iowa	Full approval 07/14/97	(62 *FR* 37514)
Kansas	Full approval 01/30/96	(61 *FR* 2938)
Kentucky	Interim approval[1] 11/14/95	(60 *FR* 57186)
Louisville/Jefferson County	Full approval 03/22/96	(61 *FR* 11738)
Louisiana	Full approval 09/12/95	(60 *FR* 47296)
Maine	Interim approval[1] 02/21/97	(62 *FR* 7939)
Maryland	Interim approval 07/03/96	(61 *FR* 34733)
Massachusetts	Interim approval 05/15/96	(61 *FR* 24460)
Michigan	Interim approval 06/18/97	(62 *FR* 33010)
Minnesota	Interim approval[1] 06/16/95	(60 *FR* 31637)
Mississippi	Full approval 12/28/94	(59 *FR* 66737)
Missouri	Full approval 05/14/97	(62 *FR* 26405)
Montana	Interim approval 05/11/95	(60 *FR* 25143)
Nebraska	Full approval 10/18/95	(60 *FR* 53872)
Lincoln–Lancaster County	Same as above	Same as above
Omaha–Douglas County	Same as above	Same as above
Nevada	Interim approval 12/12/95	(60 *FR* 63631)
Clark County	Interim approval 07/13/95	(60 *FR* 36070)
Washoe County	Interim approval 01/05/95	(60 *FR* 1741)
New Hampshire	Interim approval 10/02/96	(61 *FR* 51370)
New Jersey	Interim approval 05/16/96	(61 *FR* 24715)
New Mexico	Full approval 11/26/96	(61 *FR* 60032)
Albuquerque	Same as above	Same as above
New York	Interim approval 11/07/96	(61 *FR* 57589)
North Carolina	Interim approval 11/15/95	(60 *FR* 57357)
Forsyth County	Same as above	Same as above
Mecklenburg County	Same as above	Same as above
Western North Carolina	Same as above	Same as above
North Dakota	Interim approval 07/07/95	(60 *FR* 35335)
Northern Marianas Islands	Conditional exemption 11/13/96	(61 *FR* 58284)
Ohio	Full approval 08/15/95	(60 *FR* 42045)

TABLE 6-6
Status of State and Local Operating Permit Programs—Continued

State/local agency	*Final action*	*FR reference*
Oklahoma	Interim approval[1] 02/05/96	(61 *FR* 4220)
Oregon	Full approval 09/28/95	(60 *FR* 50106)
Lane Regional	Same as above	Same as above
Pennsylvania	Full approval 07/30/96	(61 *FR* 39597)
Puerto Rico	Full approval 02/26/96	(61 *FR* 7073)
Rhode Island	Interim approval[1] 05/06/96	(61 *FR* 20150)
South Carolina	Full approval 06/26/95	(60 *FR* 32913)
South Dakota	Full approval 01/29/96	(61 *FR* 2720)
Tennessee	Interim approval 07/29/96	(61 *FR* 39335)
Davidson County	Full approval 02/14/96	(61 *FR* 5705)
Hamilton County	Full approval 03/26/96	(61 *FR* 13101)
Knox County	Full approval 04/30/96	(61 *FR* 18966)
Shelby County	Interim approval 07/29/96	(61 *FR* 39335)
Texas	Interim approval[1] 06/25/96	(61 *FR* 32693)
Utah	Full approval 06/08/95	(60 *FR* 30192)
Vermont	Interim approval 10/02/96	(61 *FR* 51368)
Virgin Islands	Interim approval 07/31/96	(61 *FR* 39882)
Virginia	Interim approval 06/10/97	(62 *FR* 31516)
Washington	Interim approval 12/08/95	(60 *FR* 62992)
Benton Franklin County	Same as above	Same as above
Northwest	Same as above	Same as above
Olympic	Same as above	Same as above
Puget Sound	Same as above	Same as above
Southwest	Same as above	Same as above
Spokane County	Same as above	Same as above
Washington–Energy Facility Site Evaluation Council	Same as above	Same as above
Yakima County	Same as above	Same as above
West Virginia	Interim approval 11/15/95	(60 *FR* 57352)
Wisconsin	Interim approval 03/06/95	(60 *FR* 12128)
Wyoming	Interim approval 01/19/95	(60 *FR* 3766)

[1]Source category-limited interim approval.

Source: Compiled by Elsevier Science Inc.

Additional References

Additional operating permit program details can be obtained by consulting the following articles in previous issues of *The Air Pollution Consultant*:

- Page 2.1 of the September/October 1992 issue contains a review of the original 40 *CFR* Part 70 rulemaking.
- The entire August/September 1994 issue, titled *The CAA Operating Permit Program: A Guide to Compliance, 1994 Edition*, presents an in-depth examination of the Title V operating permit program.
- Page 2.1 of the November/December 1994 issue contains a review of EPA's first attempt to revise Part 70 regulations (August 29, 1994; 59 *FR* 44460). In this action, the agency proposed to 1) add a fourth permit revision track called the *de minimis* track, 2) revise the criteria for applying the now four revision tracks, and 3) implement various other changes to Part 70 rules.
- Page 2.1 of the November/December 1995 issue reviews EPA's second proposal to revise Part 70 rules (August 31, 1995; 60 *FR* 45530). This action essentially scrapped the roundly criticized August 1994 proposal in favor of a scheme to allow states and localities much greater flexibility in defining permit revision requirements subject to minimum Part 70 criteria.
- Page 2.31 of the November/December 1995 issue summarizes EPA's "White Paper for Streamlined Development of Part 70 Permit Applications."
- Page 2.15 of the July/August 1996 issue summarizes EPA's "White Paper Number 2."
- Page 4.1 of the November/December 1996 issue provides a review of the federal operating permit program regulations codified in 40 *CFR* Part 71.
- Page 2.9 of the July/August 1998 issue describes EPA's March 25, 1998 release of revised versions of certain noncontroversial sections of Part 70 (63 *FR* 14392). (Note that the agency had released a complete new draft final version of Part 70 rules in early June 1997, but withdrew this action due to renewed criticism. Re-release of revised draft final versions of the controversial sections of Part 70 that deal with permit contents and permit revisions is now expected sometime "in the future.")
- Page 2.5 of the November/December 1998 issue describes EPA's latest extension of interim approval expiration dates in light of continued problems the agency is having promulgating revised Part 70 regulations.

Chapter 7 of Elsevier Science Inc.'s 1995 publication, *Complying With Clean Air Act Regulations: Issues and Techniques*, also describes operating permit program requirements. APC

7 The Stratospheric Ozone Protection Program

When Congress amended the CAA in 1990, it directed EPA to implement the terms of the Montreal Protocol (an international agreement to protect stratospheric ozone) through the Title VI stratospheric ozone protection program. The resulting regulations, which are codified in 40 *CFR* Part 82, are designed to phase out essentially all ozone-depleting substances (ODS) by prohibiting production of such materials. The Part 82 program initially targets the most threatening ODS (i.e., Class I materials—those with the highest ozone depletion potential); other compounds (i.e., Class II ODS) are slated for less rapid phaseout.

For the interim period while the phaseout is being implemented, Part 82 specifies a variety of controls on ODS usage and handling designed to extend current supplies and allow more time for alternatives to be developed. A list of ODS appears in Table 7-1. A summary of sources subject to Part 82 requirements, along with a brief synopsis of the applicable provisions, is presented in Table 7-2 (page 7.2).

Class I and II Substances

EPA regulates Class I and II substances under the production phaseout requirements. As shown in Table 7-1, Class I compounds are divided into seven groups:

- *Group I*—Certain commonly used fully halogenated chlorofluorocarbons (CFCs) that are found in commercial products, such as refrigerants, solvents, and foams;

TABLE 7-1
Ozone-Depleting Controlled Substances

Class I Substances	All isomers of these chemicals	HCFC-142b
Group I	*Group IV*	HCFC-221
CFC-11	Carbon tetrachloride	HCFC-222
CFC-12	*Group V*	HCFC-223
CFC-113	Methyl chloroform	HCFC-224
CFC-114	All isomers of this chemical except 1,1,2-trichloroethane	HCFC-225ca
CFC-115	*Group VI*	HCFC-225cb
All isomers of these chemicals	Methyl bromide	HCFC-226
Group II	*Group VII*	HCFC-231
Halon-1211	Hydrobromofluorocarbons	HCFC-232
Halon-1301		HCFC-233
Halon-2402	**Class II Substances**	HCFC-234
All isomers of these chemicals	HCFC-21	HCFC-235
Group III	HCFC-22	HCFC-241
CFC-13	HCFC-31	HCFC-242
CFC-111	HCFC-121	HCFC-243
CFC-112	HCFC-122	HCFC-244
CFC-211	HCFC-123	HCFC-251
CFC-212	HCFC-124	HCFC-252
CFC-213	HCFC-131	HCFC-253
CFC-214	HCFC-132b	HCFC-261
CFC-215	HCFC-133a	HCFC-262
CFC-216	HCFC-141b	HCFC-271
CFC-217		All isomers of these chemicals

CFC = chlorofluorocarbon; HCFC = hydrochlorofluorocarbon.

Source: 40 *CFR* Part 82, Subpart A, Appendixes A and B.

- *Group II*—Certain halons (brominated compounds used mainly as fire extinguishing chemicals);
- *Group III*—All other fully halogenated CFCs that are not listed as Group I substances;
- *Group IV*—Carbon tetrachloride (a solvent, grain fumigant, and feedstock for CFC production);
- *Group V*—Methyl chloroform (also known as 1,1,1-trichloroethane, a widely used industrial solvent);
- *Group VI*—Methyl bromide; and
- *Group VII*—Hydrobromofluorocarbons (HBFCs).

Class II contains hydrochlorofluorocarbons (HCFCs) and their isomers. While HCFCs are also considered a threat to stratospheric ozone, they do not deplete ozone as rapidly as Class I substances. Thus, they are not as stringently regulated. In fact, many sources have taken the interim step of replacing a Class I ODS in a given usage with a Class II alternative. Eventually, a second, non-ozone-depleting alternative will be required.

Phaseout Schedules

The phaseout schedule for Class I substances is listed in Table 7-3. According to the schedule, routine production and consumption of all Class I, Group I–V compounds and HBFCs (Group VII) were banned as of 1996. Due to a lack of suitable alternatives, a complete phaseout of methyl bromide (Group VI) will not take effect until January 1, 2001. In addition, continued (albeit greatly limited) production of Class I ODS is allowed 1) for export to so-called Article 5

TABLE 7-2

Sources Subject to the Stratospheric Ozone Protection Program

- All producers and importers of ozone-depleting chemicals are subject to an allowance system
- All entities using transportation and storage containers for Class I or II substances and users of products containing or manufactured with Class I substances must meet labeling requirements[1]
- Persons manufacturing, servicing, repairing, or disposing appliances must comply with recycling and recovery standards, safe disposal practices, and the venting prohibition for Class I or II substances
- Persons servicing motor vehicle air conditioners must follow certain work practices and equipment standards
- Entities producing or distributing nonessential products (those containing ozone-depleting substances for uses that have been deemed to be nonessential) were required to cease production and distribution of these products

[1]Beginning January 1, 2015, labeling provisions apply to products containing or manufactured with Class II substances.

Source: Elsevier Science Inc.

TABLE 7-3

Phaseout Schedule for Class I Controlled Substances, Groups I–VII

Year[1]	Allowed production (% of 1986 baseline)					
	Groups I and III (CFCs)	*Group II (halons)*	*Group IV (carbon tetrachloride)*	*Group V (methyl chloroform)*	*Group VI (methyl bromide)*	*Group VII (HBFCs)*
1994	25	0	50	50	100	100
1995	25	0	15	30	100	100
1996	0	0	0	0	100	0
1997–2000	0	0	0	0	100	0
2001	0	0	0	0	0	0

CFC = chlorofluorocarbon; HBFC = hydrobromofluorocarbon.

[1]Beginning January 1.

Source: Adapted from 40 *CFR* §82.4.

TABLE 7-4
Accelerated Phaseout Schedule for Certain Class II Controlled Substances

Date	*Affected compound(s)*	*Restriction imposed by EPA*
January 1, 2003	HCFC-141b	Ban on production and consumption
January 1, 2010	HCFC-22 and HCFC-142b	Production and consumption frozen at baseline levels; ban on use of virgin chemical unless used as feedstock or for the purpose of servicing equipment manufactured before January 1, 2010
January 1, 2015	All other HCFCs[1]	Production and consumption frozen at baseline levels; ban on use of virgin chemical unless used as feedstock or as refrigerant in equipment manufactured before January 1, 2020
January 1, 2020	HCFC-22 and HCFC-142b	Ban on production and consumption
January 1, 2030	All other HCFCs[1]	Ban on production and consumption

HCFC = hydrochlorofluorocarbon.

[1]"All other HCFCs" include all listed Class II HCFCs except HCFC-22, HCFC-141b, and HCFC-142b.

Source: Elsevier Science Inc.; adapted from 40 *CFR* §82.4.

developing countries, 2) for approved "essential uses," and 3) to offset credits received for previous destruction or transformation of other Class I ODS.

For Class II ODS, see table 7-1, the three HCFCs, having the highest ozone depletion potentials (i.e., HCFC-22, HCFC-141b, and HCFC-142b) are being phased out first. As shown in Table 7-4, all other Class II substances will be phased out beginning in 2015.

Regulatory Compliance Calendar

Regulatory compliance deadlines under the stratospheric ozone protection program are compiled in Table 7-5 (pages 7.4–7.6).

Additional References

Other *APC* articles providing information on the stratospheric ozone protection program include the following:

- Page 2.1 of the November/December 1992 issue, which reviews the original Part 82 regulations promulgated on July 30, 1992 (57 *FR* 33754).
- Page 2.1 of the July/August 1994 issue, which describes the significant new alternatives policy (SNAP) program finalized on March 18, 1994 (59 *FR* 13044).

  (The SNAP program is designed to ensure that new substances developed to replace ODS do not themselves pose environmental or safety hazards. The referenced article contains the original list of SNAP-approved ODS alternatives along with a description of SNAP program rules. EPA has since issued a number of rulemakings adding to the SNAP program list; readers are advised to see the *APC* cumulative index for a listing of associated review articles.)
- Page 2.17 of the September/October 1995 issue, which details a new streamlined allowance system (May 10, 1995; 60 *FR* 24970) for Class I ODS production and use that EPA implemented in conjunction with the near-complete phaseout of production of these materials that took effect at the end of 1995.
- Page 4.18 of the March/April 1996 issue, which contains a useful overview of the Title VI stratospheric ozone protection program.
- Page 2.2 of the September/October 1998 issue, which describes EPA's June 11, 1998 proposal to expand the application of Part 82 venting controls and other prohibitions to ODS substitutes (63 *FR* 32044). APC

TABLE 7-5
Stratospheric Ozone Protection Program Regulatory Compliance Calendar[1]

Compliance date[2]		Applicability and regulatory requirements[2]	CFR reference (Title 40)	FR reference	APC review[3]
Existing sources	New sources				
Nonessential ODS usage ban					
02/16/93	Same as existing sources	The use of nonessential CFC-containing party streamers, noise horns, and alarms and the nonessential, noncommercial use of CFCs as electronic and photographic equipment cleaning fluids is banned.	Part 82, Subpart C	01/15/93 58 *FR* 4768	3:3, 2.22 (Class I ODS ban)[4]
01/17/94	Same as existing sources	The nonessential use of CFCs is banned in 1) the manufacture of flexible plastic and packaging foam products listed in §82.66(c), and 2) other aerosol and pressurized dispenser applications.			
01/01/94	Same as existing sources	The following uses of Class II ODS are banned: 1) nonessential aerosol and pressurized dispenser applications, 2) applications involving the manufacture of any plastic foam product except those exceptions listed in §82.70(c)(2), and 3) nonessential, noncommercial cleaning of electronic and photographic equipment.		12/30/93 58 *FR* 69638	
01/01/96	Same as existing sources	Ban on the use of integral skin foam that provides for motor vehicle safety where such foam is manufactured after the compliance date.			
		Ban on HCFC use in portable fire extinguishers eased.		12/04/96 61 *FR* 64424	7:2, 2.38
ODS labeling					
Later of one year after designation of substance as an ODS or 05/15/93	Same as existing sources	All products containing or manufactured using a Class I ODS and any container used to store an ODS must bear a warning label meeting the requirements of Part 82, Subpart E.	Part 82, Subpart E	02/11/93 58 *FR* 8136 01/19/95 60 *FR* 4010	3:3, 2.11 (original rule) 5:3, 2.30 (exemptions)
Later of one year after determination that a substitute exists or 01/01/15	Same as existing sources	All products containing or manufactured using a Class II ODS, where a non-ODS substitute is determined to exist, must bear a warning label meeting the requirements of Part 82, Subpart E.			

TABLE 7-5
Stratospheric Ozone Protection Program Regulatory Compliance Calendar[1]—Continued

Compliance date[2]		Applicability and regulatory requirements[2]	CFR reference (Title 40)	FR reference	APC review[3]
Existing sources	*New sources*				
Non-MVAC appliance servicing[5]					
06/14/93	Same as existing sources	Venting of ODS to the atmosphere during service, repair, or disposal of non-MVAC ODS-bearing appliances is prohibited.	Part 82, Subpart F	05/14/93 58 *FR* 28660	3:5, 2.26[4]
07/13/93	Same as existing sources	The service, repair, or disposal of non-MVAC ODS-bearing appliances must be conducted according to the refrigerant recovery standards of §82.156 using equipment meeting the requirements of §82.158.			
08/12/93	Same as existing sources	The servicing of non-MVAC appliances may only be conducted by technicians certified per §82.162.			
MVAC servicing					
06/14/93	Same as existing sources	Venting of ODS to the atmosphere during service, repair, or disposal of MVACs is prohibited.	Part 82, Subpart B[6]	07/14/92 57 *FR* 31242	2:6, 2.28[4]
08/13/92	Same as existing sources	The servicing of MVACs may only be conducted by technicians certified per §82.40 using equipment meeting requirements of §82.36.			
SNAP program					
04/18/94	Same as existing sources	Materials purchased after March 18, 1994 may be used as substitutes for ODS only if listed on the SNAP program approved list.	Part 82, Subpart G	03/18/94 59 *FR* 13044	4:4, 2.1[7]
ODS production					
01/01/96 (all Class I ODS except methyl bromide)	Same as existing sources	Normal production, importation, and consumption of ODS must cease on the listed compliance deadline; exceptions are provided for approved essential uses, production and export to developing countries, and certain other circumstances.	Part 82, Subpart A	07/30/92 57 *FR* 33754	2:6, 2.1 (original program)
01/01/01 (methyl bromide)				12/10/93 58 *FR* 65018	4:2, 2.36 (accelerated phaseout)
01/01/03 (HCFC-141b)				05/10/95 60 *FR* 24970	5:5, 2.17 (streamlined controls)[4]
01/01/20 (HCFC-22 and HCFC-142b)					
01/01/30 (all other HCFCs)					

TABLE 7-5
Stratospheric Ozone Protection Program Regulatory Compliance Calendar[1]—Continued

Compliance date[2]		*Applicability and regulatory requirements*[2]	*CFR reference (Title 40)*	*FR reference*	*APC review*[3]
Existing sources	*New sources*				
Halon blends and halon-bearing equipment					
04/06/98	Same as existing sources	Except for aviation fire protection, the manufacture of halon blends is prohibited. Halon-bearing equipment may only be disposed by manufacturers meeting terms of §82.270(d). Halon may not be disposed; it must be sent to a firm recycling the halon or destroying the material using a technology listed in §82.270(e). Venting of halon during testing, maintenance, service, or repair of halon-bearing equipment is prohibited.	Part 82 Subpart H	03/05/98 (63 *FR* 11084)	8:4, 2.40
09/01/98	Within 30 days of technician hiring, or 09/01/98, whichever is later	The servicing of halon-bearing equipment may only be conducted by technicians certified per the terms of §82.270(c).			

APC = *The Air Pollution Consultant*; CFC = chlorofluorocarbon; HCFC = hydrochlorofluorocarbon; MVAC = motor vehicle air conditioner; ODS = ozone-depleting substance; RACT = reasonably available control technology; SNAP = significant new alternatives policy.

[1]The calendar does *not* include rules and programs in the following five areas: 1) programs that are implemented on a state or local level where compliance deadlines are based on the individual agency's action (e.g., Title V operating permit programs and RACT requirements for sources in ozone nonattainment areas); 2) rules that impose deadlines on state or local government agencies (e.g., 40 *CFR* Part 58 regulations that require states and localities to establish ambient air quality monitoring systems); 3) rules promulgated prior to May 8, 1991; 4) rules that affect mobile sources; and 5) proposed rules that have not yet been promulgated.

[2]Various exemptions, extensions, and/or exceptions apply for some rules. These are *not* listed in this table. Also not listed are compliance dates for sources that exceed applicability thresholds after the promulgation of a given rule. Check the referenced *APC* review article or the listed section of the *Code of Federal Regulations* for details on these issues.

[3]Refers to the volume, issue, and page number of applicable *APC* review article. For example, 3:3, 2.22 refers to the *APC* article beginning on page 2.22 of Volume 3, Issue 3.

[4]Several other rulemakings affecting this program have been proposed or finalized since the original rule; these are *not* listed in the table. Refer to the *APC* cumulative index for a listing of review articles for these actions.

[5]The major deadlines are included; however, other deadlines exist. See Part 82, Subpart F for details.

[6]The ban on venting ODS to the atmosphere during MVAC service, repair, or disposal is actually set forth in §82.154 (of Part 82, Subpart F).

[7]Numerous rulemakings adding to the list of acceptable and unacceptable ODS substitutes have been published subsequent to the initial SNAP program promulgation. See the *APC* cumulative index for a listing of review articles for these actions.

Source: Elsevier Science Inc.

8 The Acid Rain Program

Congress created the Title IV acid rain provisions to limit SO_2 and NO_x emissions, both of which are acid rain precursors. Fossil fuel-fired electric utilities emit the majority of these pollutants and therefore are the only industry targeted for regulation under this program. Key acid rain regulations are listed in Table 8-1. A logic diagram that can be used to determine applicability of the acid rain program appears in Figures 8-1(a) and 8-1(b) (pages 8.2 and 8.3, respectively).

Table 8-2 (page 8.4) is an overview of the SO_2 emission reduction program. EPA uses a unique marketable allowance system to limit SO_2 emissions from affected sources. Each allowance permits a utility to emit one ton of SO_2 during one year. EPA controls the number of allowances issued each year and allocates allowances to regulated utilities. Allowances may be used during the year for which they were issued or during any subsequent year.

The 1990 amendments specify NO_x emission limits for two categories of boilers: Group 1 (tangentially fired and dry-bottom, wall-fired boilers) and Group 2 (cell-burner boilers, certain cyclone boilers, certain wet-bottom boilers, and vertically fired boilers). These emission limits and their associated compliance deadlines are listed in Table 8-3 (page 8.4). Note that the Group 1 boilers that are listed as Phase I units under 40 *CFR* §73.10 must comply with emission limits earlier than the other Group 1 boilers (i.e., by January 1, 1996), but are *not* then subject to the more stringent Phase II limits, which take effect for most sources on January 1, 2000. Acid rain NO_x limitation applicability criteria and compliance dates are summarized in Figure 8-2 (page 8.5).

A regulatory compliance calendar for the acid rain program appears in Table 8-4 (page 8.6).

For further information on the acid rain program, refer to the following articles in *APC:*

- Page 2.1 of the March/April 1993 issue, which reviews the original acid rain program rules issued on January 11, 1993 (58 *FR* 3590).
- Page 2.12 of the July/August 1995 issue, which describes the acid rain opt-in program under which sources may voluntarily subject themselves to SO_2 limits in return for receiving marketable SO_2 allowances.
- Page 4.10 of the March/April 1996 issue, which contains a brief synopsis of acid rain program rules.

TABLE 8-1
Key Acid Rain Regulations Under 40 *CFR* Parts 72–78

Part 72	Core acid rain program regulations
Part 73	SO_2 marketable allowance system
Part 74	Opt-in provisions
Part 75	Monitoring requirements
Part 76	Requirements for NO_x reductions
Part 77	Regulations governing excess emissions
Part 78	Administrative appeal provisions

NO_x = nitrogen oxides; SO_2 = sulfur dioxide.

Source: Elsevier Science Inc.

Besides the highlight articles listed above, EPA has issued a number of other less important acid rain program rulemakings, all of which have been reviewed in *APC*. Interested readers should see the cumulative index of the journal for details.

Finally, note that certain electric utility boilers are also subject to emission controls under other CAA programs. Two such programs that have seen recent regulatory action are as follows:

- NSPS codified at 40 *CFR* Part 60, Subpart Da and revised on September 16, 1998 (63 *FR* 49442; see the review article that will appear in the January/February 1999 issue of *APC*). This NSPS applies stringent new fuel-neutral NO_x limits to utility boilers constructed, reconstructed, or modified beginning after July 9, 1997.
- Proposed NO_x emission reduction requirements that will be imposed on most states east of the Mississippi river to address long-range NO_x transport and downrange ground-level ozone formation. When promulgated, the regional NO_x transport rule will undoubtedly spawn either 1) additional NO_x limits applicable to existing and new fossil fuel-fired utility boilers, and/or 2) new NO_x cap-and-trade programs similar to the existing acid rain program SO_2 allowance system. (The latest proposed rule on this subject was published on May 11, 1998 [63 *FR* 25902]; see the review article on page 2.12 of the September/October 1998 issue of *APC*.)

FIGURE 8-1(a)
Determining Applicability of the Acid Rain Program

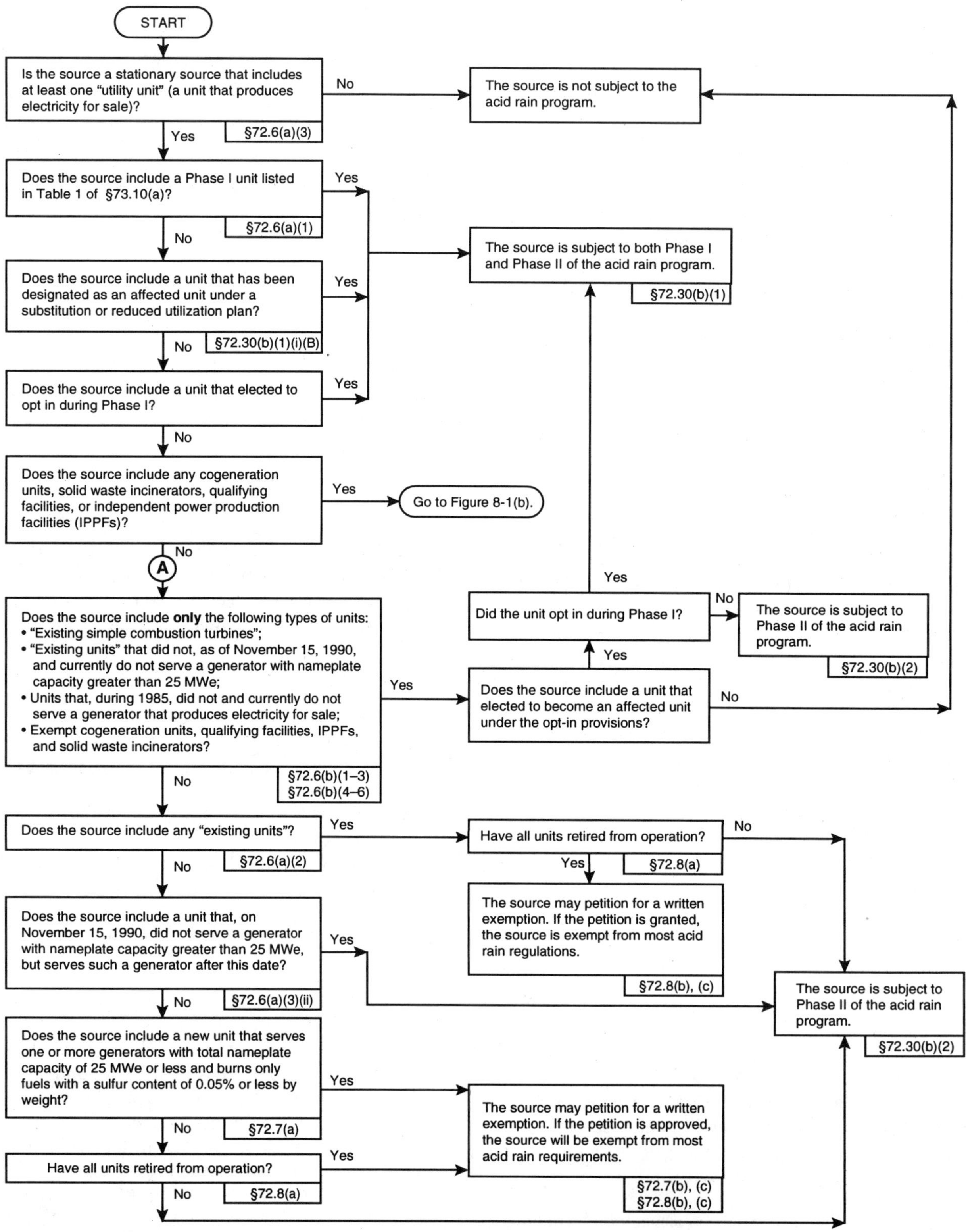

Source: Elsevier Science Inc.

FIGURE 8-1(b)
Determining Applicability of the Acid Rain Program—Continued

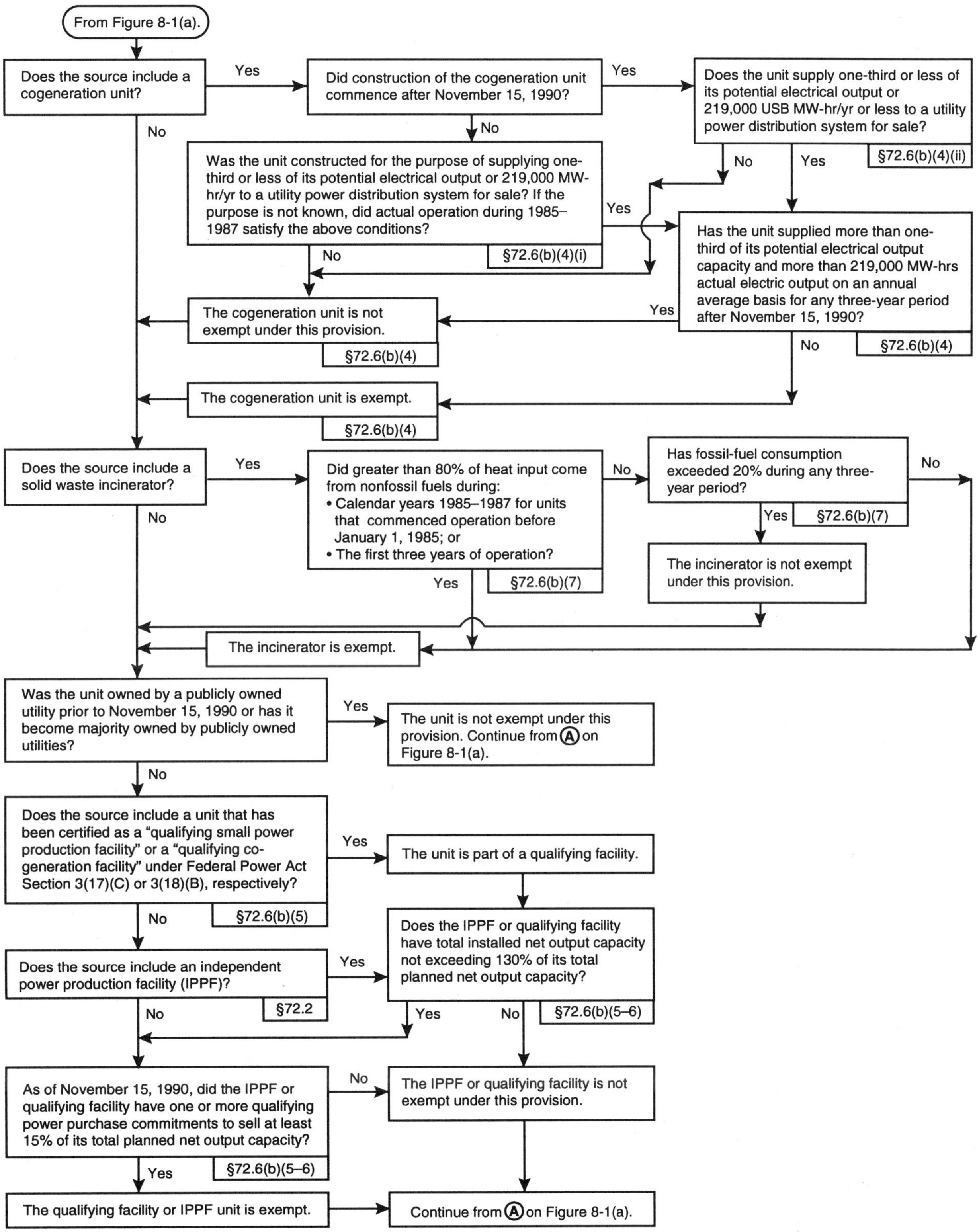

Source: Elsevier Science Inc.

TABLE 8-2
Overview of the SO_2 Emission Reduction Program

	Summary	*Effective date*
Phase I	Limits SO_2 emissions from the 110 highest polluting units in the nation (listed in Table 1 of 40 *CFR* §73.10) by allocating a certain number of SO_2 allowances to each of these facilities. Collectively, the allocated Phase I allowances are designed to limit SO_2 emissions from these units to 2.50 lb/MMBtu of baseline fuel consumption (i.e., generally the annual average fuel consumption for 1985, 1986, and 1987).	January 1, 1995
Phase II	Will regulate virtually all electric utility units that produce at least 25 megawatts of electricity. In Phase II, total allocated SO_2 emissions will generally be limited to 8.95 million tons (which is equivalent to about 1.20 lb/MMBtu of baseline fuel consumption).	January 1, 2000

SO_2 = sulfur dioxide.

Source: Elsevier Science Inc.

TABLE 8-3
Acid Rain Program NO_x Emission Limits

Boiler type	*Emission limit (lb NO_x/MMBtu of heat input)*	*Compliance deadline*
Group 1		
Phase I		
Tangentially fired boilers	0.45	January 1, 1996
Dry-bottom, wall-fired boilers	0.50	Same
Phase II		
Tangentially fired boilers	0.40	January 1, 2000[a]
Dry-bottom, wall-fired boilers	0.46	Same[a]
Group 2		
Cell-burner boilers	0.68	January 1, 2000
Cyclone boilers with 100% load steam flow of $>1.06 \times 10^6$ lb/hr	0.86	Same
Wet-bottom boilers with 100% load steam flow of $>0.45 \times 10^6$ lb/hr	0.84	Same
Vertically fired boilers	0.80	Same

NO_x = nitrogen oxides.

[a]Group 1, Phase II units that have voluntarily elected to meet the Phase I limits (i.e., early electing units) have until January 1, 2008 to meet Phase II limits.

Source: Elsevier Science Inc.; compiled from §§76.5(a), 76.6(a), 76.7(a), and 76.8.

FIGURE 8-2

Applicability Criteria and Compliance Dates for Acid Rain NO_x Limitations

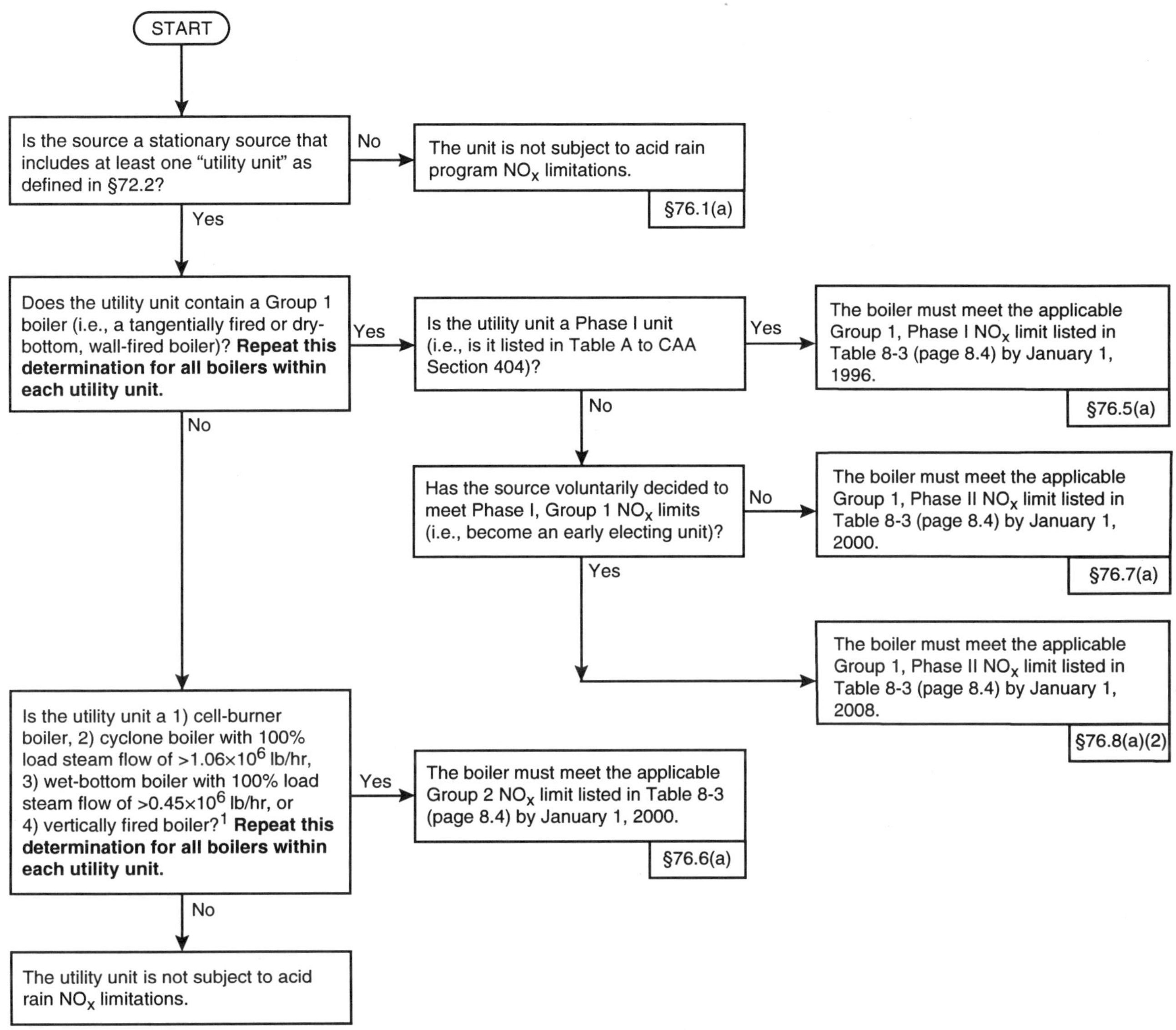

NO_x = nitrogen oxides.

[1]Note that NO_x limitations are not imposed on all types of Group 2 boilers (e.g., fluidized-bed boilers are not regulated).

Source: Elsevier Science Inc.

TABLE 8-4
Acid Rain Program Regulatory Compliance Calendar[1]

Compliance date[2]: Existing sources	Compliance date[2]: New sources	Applicability and regulatory requirements[2]	CFR reference (Title 40)	FR reference	APC review[3]
SO_2 emission limits					
01/01/95 (Phase I)	Same as existing sources	Units at 110 highest emitting utility plants must limit SO_2 emissions to below 2.50 lb/MMBtu. Limits are imposed through a system of marketable allowances, where each allowance entitles the holder to emit 1 ton of SO_2 per year. Program includes requirement to obtain an acid rain permit.	Parts 72, 73, 75, and 77	01/11/93 58 *FR* 3590	3:2, 2.1[4]
01/01/00 (Phase II)	Same as existing sources	Most electric utility units with output greater than 25 MWe electricity must limit SO_2 emissions to below 1.20 lb/MMBtu. As with Phase I, limits are imposed through a marketable allowance system. Program also includes permits.			
NO_x emission limits					
01/01/96 (Group 1, Phase I)	Same as existing sources	Tangentially fired and dry-bottom, wall-fired Phase I boilers (i.e., Group 1, Phase I boilers) must meet NO_x limits of 0.45 lb NO_x/MMBtu and 0.50 lb NO_x/MMBtu, respectively.	Part 76	03/22/94 59 *FR* 13538 04/13/95 60 *FR* 18751	4:3, 2.1 (Group 1, Phase I limits) 5:4, 2.7 (Group 1, Phase I compliance extension)
01/01/00 (Group 1, Phase II)	Same as existing sources	Tangentially fired and dry-bottom, wall-fired Phase II boilers (i.e., Group 1, Phase I boilers) must meet NO_x limits of 0.40 lb NO_x/MMBtu and 0.46 lb NO_x/MMBtu, respectively.		12/19/96 61 *FR* 67112	7:2, 2.9 (Phase II limits)
01/01/00 (Group 2, Phase II)	Same as existing sources	Group 2 boilers must meet the following Phase II NO_x limits: cell-burner boilers, 0.68 lb NO_x/MMBtu; cyclone burners above 155 MWe size, 0.86 NO_x/MMBtu; wet-bottom boilers above 65 MWe size, 0.84 NO_x/MMBtu; and vertically fired boilers, 0.80 NO_x/MMBtu. (There are no Phase I limits for Group 2 boilers.)		06/12/97 62 *FR* 30238	7:5, 2.14 (Correction)

APC = *The Air Pollution Consultant*; NO_x = nitrogen oxides; RACT = reasonably available control technology; SO_2 = sulfur dioxide.

[1]The calendar does *not* include rules and programs in the following five areas: 1) programs that are implemented on a state or local level where compliance deadlines are based on the individual agency's action (e.g., Title V operating permit programs and RACT requirements for sources in ozone nonattainment areas); 2) rules that impose deadlines on state or local government agencies (e.g., 40 *CFR* Part 58 regulations that require states and localities to establish ambient air quality monitoring systems); 3) rules promulgated prior to May 8, 1991; 4) rules that affect mobile sources; and 5) proposed rules that have not yet been promulgated.

[2]Various exemptions, extensions, and/or exceptions apply for some rules. These are *not* listed in this table.

[3]Refers to the volume, issue, and page number of applicable *APC* review article. For example, 3:2, 2.1 refers to the *APC* article beginning on page 2.1 of Volume 3, Issue 2.

[4]Several other rulemakings affecting this program have been proposed or finalized since the original rule; these are *not* listed in the table. Readers should see the *APC* cumulative index for a listing of review articles for these actions.

Source: Elsevier Science Inc.

9 Where to Go for More Information

In this chapter, we provide detailed instructions for locating various regulatory documents, models, technical guidance, and other key sources of CAA information. Due to the recent surge of online CAA information, we include Internet resources whenever possible. (The Internet addresses provided in this publication were current as of October 1, 1998.) In addition to online resources, traditional methods of information access (obtaining print copies of publications, contacting hotlines, etc.) are also mentioned. The following topics are covered in this chapter:

A. CAA References

Although most of us work closely with the CAA regulations on a day-to-day basis, sometimes we also need to consult the statute. The full text of the CAA and the 1990 amendments can be obtained from EPA's Web site at http://www.epa.gov/oar/oaq_caa.html. The agency has also created *The Plain English Guide to the Clean Air Act*, which explains the statute in easy-to-understand terms. It can be accessed via http://www.epa.gov/oar/oaqps/peg_caa/pegcaain.html#index.

The *United States Code*, which contains all of the laws of the United States (including the CAA), is also available online from the Government Printing Office (GPO) through GPO Access at http://www.access.gpo.gov/su_docs/aces/aaces002.html. Information on current bills and bills recently passed by Congress (including those involving air issues) can also be obtained from GPO Access. Another comprehensive source of legislative information is the Thomas Legislative Information Web site (http://thomas.loc.gov/home/ thomas2.html).

B. The TTNWeb Site

EPA's Office of Air Quality Planning and Standards (OAQPS) operates one of the most comprehensive CAA Web sites—the technology transfer network (TTN) Web site. The TTNWeb (http://www.epa.gov/ttn) is a collection of 16 Web sites that contain information about many areas of air pollution regulation, technology, measurement, and prevention. Table 9-1 (page 9.2) is a list of the Web sites included in the TTNWeb. (Descriptions and road maps of key sites contained in the TTNWeb are also provided later in this chapter and in Chapter 10, respectively.)

Through the TTNWeb, users can obtain numerous technical and regulatory guidance documents, retrieve newly issued CAA regulations (usually prior to their publication in the *FR*), download tools to estimate air pollutant emissions, locate training information, request technical support, and participate in open forums.

TABLE 9-1
Directory of Sites in the TTNWeb[1]

TTNWeb site	Summary of contents	URL
AIRS—Aerometric Information Retrieval System	Access to the AIRS computer database, which contains air emission data from individual sources and monitoring stations across the United States.[2] Turn to page 10.3 for a road map of the AIRS Web site.	http://www.epa.gov/airs
AMTIC—Ambient Monitoring Technology Information Center	Information on ambient air monitoring programs, details on monitoring methods, relevant documents and articles, information on air quality trends and nonattainment areas, and federal ambient air quality monitoring regulations. A road map of AMTIC appears on page 10.5.	http://www.epa.gov/ttn/amtic
CHIEF—Clearinghouse for Inventories and Emission Factors	EPA's latest information and tools for estimating air pollutant emissions and performing air emission inventories. See page 10.7 for the CHIEF road map.	http://www.epa.gov/ttn/chief
CATC—Clean Air Technology Center	Technical resources on emerging and existing air pollution prevention and control technologies, including: • Access to the RACT/BACT/LAER Clearinghouse (RBLC), which contains data on prevention and control technology determinations made by states and local permitting agencies; this information can be used to help identify appropriate technologies for mitigating or treating most air pollutant emissions; and • Technical reports, cost information, and software. Turn to page 10.9 for a road map of CATC.	http://www.epa.gov/ttn/catc
CICA—U.S./Mexico Border Information Center on Air Pollution	Technical information for evaluating air pollution along the Mexico/U.S. border, including: • Border air quality data • Summaries of border air quality programs, and • Descriptions of ongoing and completed CICA projects.	http://www.epa.gov/ttn/catc/cica
EMC—Emission Measurement Center	Guidance on emission testing and monitoring methods. A road map of the EMC Web site can be found on page 10.11.	http://www.epa.gov/ttn/emc
FACA—Federal Advisory Committee Act	Information on the efforts of EPA's Clean Air Act Advisory Committee to research integrated approaches for implementing revised NAAQS for ozone and PM, as well as the regional haze program, including: • A summary of EPA's NAAQS review process, • Descriptions of regional haze/visibility issues, • Information on the health and environmental effects of ground-level ozone and PM, • A list of counties that do not meet the 8-hour ozone NAAQS, • Visibility maps, • FAQs regarding the ozone and particulate matter NAAQS, and • Access to EPA's updated NAAQS for ozone and PM.	http://www.epa.gov/ttn/faca

TABLE 9-1
Directory of in the TTNWeb[1]—Continued

TTNWeb site	*Summary of contents*	*URL*
GEI—Geographic/Ecosystem Initiative	Information on geographically-focused environmental activities, called GEIs, such as: • Ecosystems Initiatives, • The Southern Appalachian Mountains Initiative, and • The U.S./Canada Air Quality Agreement.	http://www.epa.gov/ttn/gei
ICCR—Industrial Combustion Coordinated Rulemaking	Updates on the development of the ICCR.	http://www.epa.gov/ttn/iccr/home.html
NELAC—National Environmental Laboratory Accreditation Conference	All current NELAC environmental laboratory standards, proposed changes to the standards, NELAC updates, and a directory of NELAC members.	http://www.epa.gov/ttn/nelac
NSR—New Source Review	Resources pertaining to NSR permitting, such as: • Access to current NSR rules; • Technical information on several topics, including emerging control technologies, pollution prevention projects, BACT and LAER decisions, air quality modeling, and analyses of impacts on water, soil, and vegetation; • Historical and recent letters, memoranda, and Environmental Appeals Board decisions that explain how EPA interprets the NSR provisions; • Access to nonattainment NSR and PSD memos contained in EPA's *New Source Review: Prevention of Significant Deterioration and Nonattainment Area Guidance Notebook*; and • NSR contacts.	http://www.epa.gov/ttn/nsr
OAR P&G—Office of Air & Radiation Policy & Guidance Documents	Access to recent and historical rules, policy, and guidance documents issued by OAR. Turn to page 10.17, for a road map to the OAR P&G Web site.	http://www.epa.gov/ttn/oarpg
OTAG—Ozone Transport Assessment Group	Information on OTAG's efforts to address the problem of ground-level ozone (smog) and long-range pollutant transport across the eastern United States.	http://www.epa.gov/ttn/otag
SBAP—Small Business Assistance Program	Technical assistance for small businesses that need to comply with the CAA, including: • Updates on new Internet information that will be useful for small businesses, • Fact sheets and other small business materials that explain CAA rules and requirements, • Lists of EPA and state SBAP contacts, and • Access to SBAP publications.	http://www.epa.gov/ttn/sbap

TABLE 9-1
Directory of TTNWeb Sites[1]—Continued

TTNWeb site	Summary of contents	URL
SCRAM—Support Center for Regulatory Air Models	Information on atmospheric dispersion models that support CAA regulatory programs, including: • EPA modeling project data; • Air quality dispersion models and user's guides available for downloading; • Meteorological data; • Modeling guidance, tutorials, and other useful information; and • Lists of EPA and state modeling contacts. See the SCRAM road map on page 10.18 for further information.	http://www.epa.gov/ttn/scram/main.htm
UATW—Unified Air Toxics Website	This HAP information clearinghouse includes: • EPA rules concerning HAPs (e.g., MACT, NESHAP regulations); CAA Section 112 requirements, studies and reports; HAP policy, initiatives, and guidance; • A list of HAPs; • Summaries of HAP sources; • Information on how to access technical HAP information via EPA hotlines, the Internet, and technical documents; • Overview of the CAA Section 112 air toxic provisions; and • Lists of EPA, state, and local agency contacts. A road map of the UATW can be found on page 10.19.	http://www.epa.gov/ttn/uatw

BACT = best available control technology; FAQ = frequently asked question; HAP = hazardous air pollutant; LAER = lowest achievable emission rate; MACT = maximum achievable control technology; NAAQS = national ambient air quality standards; NESHAP = national emission standards for hazardous air pollutants; PM = particulate matter; PSD = prevention of significant deterioration; RACT = reasonably available control technology; URL = uniform resource locator.

[1]The TTNWeb can be accessed via http://www.epa.gov/ttn/direct.html.

[2]As of October 1, 1998, the complete AIRS database is not publicly available via the Internet; only certain summary information is available on AIRS. However, EPA is planning to make the entire database accessible from the Internet by mid-1999.

Source: Elsevier Science Inc.

C. Regulatory Information

Due to the ever-changing nature of the regulations, anyone affected by the CAA must constantly stay in touch with the agency's regulatory actions, policies, etc. The following key regulatory information sources are outlined below: 1) the *FR*, 2) the *CFR*, 3) the TTNWeb and other EPA Web sites, 4) EPA regulatory interpretations, and 5) the Air Pollution Control Resources Web page.

The Federal Register

EPA's proposed and final rules and important agency notices are published in the *FR*. The *FR* is published daily and is available from several sources:

- *GPO Access*—The full text of *FR* documents issued by EPA and all other federal agencies from January 1994 to the present are available online from the GPO, which maintains the official electronic version of the *FR*. Using GPO Access, you can search for a particular rule or notice by date, *FR* page number, agency, *CFR* part, or keyword. The GPO Access site (http://www.access.gpo.gov/su_docs/aces/aces140.html) also provides access to the *CFR*, *Congressional Record*, and other bills, laws, and reports.

- *EPA Web site*—EPA *FR* documents issued from October 1994 to the present can also be obtained from EPA's Web site (http://www.epa.gov/fedrgstr). (The documents are downloaded from the GPO server.) The EPA *FR* Web site contains 1997 and 1998 table of contents calendars. By simply clicking on a calendar for a specific date, you can view the *FR* table of contents for that day. You can then click on any item of interest and connect directly to the full text. Current (1998) and past documents can also be located by searching by date, category (e.g., EPA-AIR, EPA-GENERAL), and/or keyword.

  If you know the date of the *FR* notice, you can jump directly to a list of air regulation documents for that day by building on the following Web address: http://www.epa.gov/fedrgstr/EPA-AIR/. To build on this Web address, add the year, the month, and the day—all separated by backslashes (/). For instance, if you want to access an *FR* document that was published on December 4, 1996, you would enter the following address in your Web browser's "location" window: http://www.epa.gov/docs/fedrgstr/EPA-AIR/1996/December/Day-04. This will take you to a list of air notices published in the *FR* on that day. From there, you can select specific notices of interest.

- *EPA listservers*—Daily *FR* tables of content and *FR* documents issued by EPA will automatically be e-mailed to anyone who subscribes to the appropriate EPA public listserver. These listservers are organized by topic, such as EPA-AIR (Office of Air and Radiation documents). A list of the available listservers and instructions on how to subscribe to these free services are available via http://www.epa.gov/epahome/listserv.htm.

- *Elsevier Science Inc.'s Air Pollution Regulatory Analysis Service*—This service offers reviews and analyses of final and proposed regulations affecting the CAA and mails these articles directly to subscribers ten working days from publication in the *FR*. Call (303) 470-1900, ext. 12 for more information.

- *Print copies*—Contact GPO at (202) 512-1800 to subscribe to the print version of the *FR*.

All EPA regulations must be reviewed and approved by the Office of Management and Budget (OMB) prior to publication in the *FR*. To get advance notice of the regulations that may be issued soon, readers can check OMB's Regulations Under Review Web page (http://library.whitehouse.gov/omb/OMBREGS.HTM#Environmental Protection Agency). The site lists regulations pending OMB review and OMB reviews completed in the last 30 days.

The Code of Federal Regulations

All CAA regulations promulgated by EPA are codified in Title 40 of the *CFR*, which is organized by part (e.g., Part 63 contains the MACT standards). A summary of the contents of each part pertaining to CAA rules is provided in Table 2-1 (page 2.2–2.12). Each volume of Title 40 is updated annually as of July 1.

The *CFR* can be accessed via:

- The GPO Access Web site (http://www.access.gpo.gov/nara/cfr/index.html), which includes all 50 titles of the *CFR*. Using GPO Access, you can search the *CFR* by keyword.

- The EPA Web site (http://www.epa.gov/epacfr40), which provides the full text of Title 40 only. The regulations are retrievable by chapter, subchapter, and part.

- GPO, which provides individual printed copies of *CFR* volumes. Copies can be ordered by calling (202) 512-1800 or through local GPO bookstores. For the GPO bookstore nearest you, call GPO or access http://www.access.gpo.gov/su_docs/sale/abkst001.html.

- Elsevier Science Inc.'s *Federal Compliance Alert Service*, which provides e-mail notification of regulatory actions on affected parts of the *CFR* specified by the subscriber. For information on this free service, see the Elsevier Science Web site at http://elsevier.enviroinfo.com.

The TTNWeb and Other EPA Web Sites

The EPA Office of Air & Radiation's Policy & Guidance Documents Web site (OAR P&G) provides easy access to regulatory and statutory information associated with the 1990 amendments. This site, which is part of the TTNWeb, offers a wide array of information and can be accessed via http://www.epa.gov/ttn/oarpg. For each section of the 1990 amendments (Titles I through XI), the OAR P&G includes CAA text and summaries, policy and guidance documents, and the status of rules and projects. Bulletins summarizing recent regulatory developments, notices of meetings and workshops, copies of recently signed rules, and recently issued policy/guidance memos are also available. The Web site is also searchable. For a useful list of the CAA policy, guidance, and regulations issued in the last six months, go to http://www.epa.gov/ttn/oarpg/ramain.html.

In addition to the OAR P&G, EPA has created other Web sites that offer comprehensive regulatory information on specific CAA programs, including:

- *Stratospheric ozone program Web page*—This site contains information about ozone depletion, EPA's stratospheric ozone regulations, substitutes for ozone-depleting substances under the SNAP program, methyl bromide, international policy regarding stratospheric ozone, resources, and EPA contacts (http://www.epa.gov/ozone).
- *Acid Rain Division's home page*—Via this Web page, the following information can be accessed: information about acid rain formation and effects, the SO_2 trading program, the NO_x reduction program, emission monitoring and reporting, pollution prevention alternatives, and EPA contacts (http://www.epa.gov/acidrain).
- *NSR Web site*—The NSR Web site, part of the TTNWeb, provides information pertaining to NSR permitting, including NSR regulations, memoranda, and guidance documents (http://www.epa.gov/ttn/nsr).
- *Unified Air Toxics Website (UATW)*—The UATW is the central clearinghouse and repository for hazardous air pollutant (air toxics) information from federal, state, and local agencies. It includes various MACT standards and NESHAPs, fact sheets on air toxics regulations, a list of technical resources, the list of HAPs, and agency contacts (http://www.epa.gov/ttn/uatw). (A road map of the UATW appears on page 10.19.)
- *Searchable EPA guidance databases*—EPA Region VII has developed two searchable databases that contain EPA policy and guidance documents for the Title V operating permit program, and the NSR and PSD programs, respectively. The operating permit database can be downloaded from http://www.epa.gov/region07/programs/artd/air/title5/titlevhp.htm#Guidance. The NSR/PSD database is available from http://www.epa.gov/region07/programs/artd/air/nsr/nsrpg.htm.

EPA Regulatory Interpretations

If you are unsure of how the CAA regulations apply to your situation or need to obtain documents, you can get assistance by 1) contacting one of the EPA hotlines listed in Table 9-2, 2) calling the appropriate EPA contact (see the EPA Contacts beginning on page 9.20), or 3) writing a letter to EPA requesting a written regulatory interpretation.

TABLE 9-2
EPA Air-Related Hotlines

Acid Rain Hotline	(202) 233-9620
Air Risk Information Center	(919) 541-0888
AIRS Hotline	(800) 334-2405
Clean Air Technology Center	(919) 541-0800
Info CHIEF Help Line	(919) 541-5285
Emission Measurement Center Source Hotline	(919) 541-0200
Indoor Air Quality Information Clearinghouse (IAQINFO)	(800) 438-4318
National Lead Information Center	(800) 424-5343
Stratospheric Ozone Hotline	(800) 296-1996
Toxics Release Inventory—User Support	(202) 260-1531

CHIEF = Clearinghouse for Information and Emission Factors.

Source: Elsevier Science Inc.

Written requests must state all of the relevant facts concerning the issue at hand and should be mailed to:

John Rasnic
Director of Manufacturing, Energy and Transportation Division (METD)
OC, Office of Enforcement and Compliance Assurance (OECA)
U.S. EPA Mail Code 2223A
401 M St. SW
Washington, DC 20460

OECA maintains a database of regulatory interpretations called the Applicability Determination Index (ADI), which is accessible via the Internet (http://ttnwww.rtpnc.epa.gov/cfdocs/adiwww/adiwww.html-ssi). The database includes EPA's responses to questions related to applicability and compliance issues associated with the NSPS, NESHAP, and stratospheric ozone regulations.

Air Pollution Control Resources Web Page

The Air Pollution Control Resources Web page provides an extensive list of federal air regulation resources (http://members.aol.com/CleenAir/apc/fed.htm#oar). This resource list, which includes nongovernment sources, was one of the best we came across during our Internet search. The Air Pollution Control Resources home page also tracks current developments regarding CAA legislation, NAAQS, hazardous air pollutant regulation, operating permits, and CAA enforcement issues (http://members.aol.com/CleenAir/index.htm). These news and development sites are gems of up-to-date information.

D. Emission Calculation Tools

The best source of current EPA emission calculation and inventory information is the Clearinghouse for Inventories and Emission Factors (CHIEF) Web site (http://www.epa.gov/ttn/chief). This site, which is compiled by EPA's Emission Factor and Inventory Group (EFIG), includes emission estimation references and software in addition to emission inventory data and guidance. The types of information available via CHIEF are outlined in Table 9-3 (pages 9.8–9.10), including locating and estimating (L&E) reports. A list of available L&E reports is provided in Table 9-4 (page 9.11). A complete list of EFIG software and documents can be viewed at http://www.epa.gov/ttn/chief/ reptdesc.html.

In addition, EFIG operates the Info CHIEF Help Line at (919) 541-5285, which offers assistance regarding emission calculation and inventory issues and publishes the CHIEF Newsletter. Copies of the newsletter can be requested from Info CHIEF. The newsletter will also be automatically e-mailed to individuals who subscribe under the EPA public listservers service. Instructions on how to subscribe to this free service are available at http://www.epa.gov/epahome/listserv.htm.

E. Dispersion Models

The Support Center for Regulatory Air Model Web page (http://www.epa.gov/ttn/scram) maintained by OAQPS is an extensive source of information on atmospheric dispersion models. From the site, files containing computer code for a wide range of screening-level and refined models can be downloaded along with user guides. The site also offers notices of model changes, meteorological data for surface and upper air readings, EPA memoranda relating to dispersion models, other EPA modeling guidance, research abstracts, and a list of state modeling contacts.

F. Monitoring Information

The TTNWeb has three sites that provide comprehensive information on monitoring methods and data: Ambient Monitoring Technology Information Center (AMTIC), Emission Measurement Technical Information Center (EMTIC), and Aerometric Information Retrieval Systems (AIRS). These resources are summarized below.

Ambient Monitoring Technology Information Center

The AMTIC Web site (http://ttnwww.rtpnc.epa.gov/html/amtic/amtic.htm#AM01), which is part of the TTNWeb, provides a wealth of information about ambient monitoring. It currently contains information on all reference methods for criteria pollutants; noncriteria pollutant methodologies will soon be included as well. Also available are federal regulations pertaining to ambient monitoring, ambient monitoring quality assurance/quality control information, publications relating to ambient monitoring, summaries of field and laboratory studies of interest, and training information.

Emission Measurement Technical Information Center

Technical guidance on stationary source emission monitoring and testing can be accessed via the TTNWeb's EMTIC Web site (http://ttnwww.rtpnc.epa.gov/html/emtic/emtic.htm#EM02). Information is available on continuous emission monitoring, compliance assurance monitoring, acid rain program monitoring, open-market trading program emission measurement, sampling and test methods, and emission testing software. Access is also provided to EMTIC guideline and information documents as well as EPA contacts.

Aerometric Information Retrieval Systems

The AIRS database contains ambient air quality, emissions, and compliance data for the United States and various World Health Organization countries. It contains air pollutant and meteorological data from about 10,000 monitoring stations, and data from almost 150,000 point sources monitored by EPA and/or state and local air regulatory agencies. A subsystem of maps and charts shows patterns, trends, and anomalies in the air pollution data.

The AIRS Web site (http://www.epa.gov/airs/airs.html) makes AIRS-related information and computerized software available to the AIRS user community. Its focus is to encourage the exchange of information among state and local agencies that use AIRS documents and information. (States use the AIRS data when preparing state implementation plans [SIPs], to track the compliance status of point sources, and report emission estimates for CAA-regulated pollutants.)

TABLE 9-3

Emission Estimation Tools Available From EPA's Clearinghouse for Inventories and Emission Factors (CHIEF)[1]

Tool	*Description*	*Source*
Air CHIEF CD-ROM	Air emission data for estimating the types and quantities of pollutants that may be emitted from a wide variety of sources including 1) *Compilation of Air Pollutant Emission Factors* (*AP-42*), 2) L&E documents, 3) the FIRE database, and 4) the TANKS database	• Available from GPO, (202) 512-1800, $15, or http://www.access.gpo.gov/su_docs/sale/prf/prf.html
Compilation of Air Pollutant Emission Factors (*AP-42*), Volume I	Process descriptions, potential air emission sources, control methods, and emission factors for over 200 stationary source categories	• http://www.epa.gov/ttn/chief/ap42.html • Print copies available from GPO, (202) 512-1800, $56 • Also included on the Air CHIEF CD-ROM • Individual sections available via automatic fax transmission from Fax CHIEF; call (919) 541-0548 or (919) 541-5626 from a touch-tone telephone
FIRE	Database containing EPA's recommended emission factors for criteria and hazardous air pollutants; provides information about various industries, including types of processes and pollutants emitted; allows for easy access to criteria air pollutant emission factors obtained from *AP-42*, hazardous air pollutant emission factors derived from source tests, and the Crosswalk/Air Toxic Emission Factor database	• http://www.epa.gov/ttn/chief/fire.html • Also included on the Air CHIEF CD-ROM
L&E documents	Detailed information on potential emission points for specific substances in the most common source categories; includes general descriptions of the emitting processes, potential emission points, and emission factors (a list of pollutants for which L&E documents have been issued is presented in Table 9-4 [page 9.11])	• http://www.epa.gov/ttn/chief/ap42etc.html#LE
MECH	Software that estimates fugitive particulate emissions from paved roads, unpaved roads, materials handling, agricultural tilling, and construction/demolition operations; automates calculations contained in Control of Open Fugitive Dust Sources (EPA 450/3-88-008)	• http://www.epa.gov/ttn/chief/software.html

TABLE 9-3
Emission Estimation Tools Available From EPA's Clearinghouse for Inventories and Emission Factors[1]—Continued

Tool	*Description*	*Source*
National inventory data and projection software	Contains Mobile Input Data Analysis System software, Ozone Transport Commission NO_x baseline inventory data, emission projection software, and national emission trend data	• http://www.epa.gov/ttn/chief/ei_data.html#psd
Landfill air emissions estimation model	Estimates methane, carbon dioxide, nonmethane organic compound, and individual toxic air pollutant emissions from landfills; determines if a landfill is subject to the NSPS for new municipal solid waste landfills (40 *CFR* Part 60, Subpart WWW) or the emission guidelines for existing municipal solid waste landfills (40 *CFR* Part 60, Subpart Cc)	• http://www.epa.gov/ttn/catc/products/html#software
MOBILE emissions model	Highway vehicle emission factor model used by state agencies and other parties to develop emission inventories and emission reduction plans required by SIPs	• http://www.epa.gov/OMSWWW/models.htm
National Emission Trends viewer CD-ROM	Summary level data from EPA's National Emission Trends database (1985–1996); sorts emissions by tier category or Standard Industrial Classification at the county, nonattainment area, state, and national levels	• Available from GPO (Stock Number: 055-000-00606-7), (202) 512-1800, $13
Personal Computer Biogenic Emissions Inventory System	Estimates hourly emissions of biogenic VOC and soil NO_x emissions for any county in the contiguous United States	• http://www.epa.gov/ttn/chief/software.html
Source Classification Codes	List of source classification codes (8-digit codes used to categorize individual processes or unit operations that generate air emissions) for emission factor purposes	• http://www.epa.gov/ttn/chief/scccodes.html
PM calculator	Estimates PM-2.5 and PM-10 control efficiencies for various pairs of particulate control devices on specific emission sources	• http://www.epa.gov/ttn/chief/software.html#pcbeis
SPECIATE	Contains organic compound and PM speciation profiles for more than 300 source types	• http://www.epa.gov/ttn/chief/software.html

TABLE 9-3
Emission Estimation Tools Available From EPA's Clearinghouse for Inventories and Emission Factors[1]—Continued

Tool	*Description*	*Source*
TANKS	Software that estimates VOC emissions from fixed- and floating-roof storage tanks; model based on *AP-42* emission estimation procedures plus recent updates from the American Petroleum Institute	• http://www.epa.gov/ttn/chief/tanks.html
WATER8	Model for estimating compound-specific air emissions from wastewater collection and treatment systems	• http://www.epa.gov/ttn/chief/software.html #water8
WIND	Software that estimates fugitive particulate emissions due to high winds; automates *AP-42* calculations for industrial wind erosion	• http://www.epa.gov/ttn/chief/software.html
CHEMDAT8	Spreadsheet that includes analytical models for estimating VOC emissions from hazardous waste treatment, storage, and disposal facility processes	• http://www.epa.gov/ttn/chief/software.html #water8

CHIEF = Clearinghouse for Inventories and Emission Factors; FIRE = Factor Information and Retrieval System; GPO = Government Printing Office; L&E = locating and estimating; NET = National Emissions Trends; NO_x = nitrogen oxides; NSPS = new source performance standards; PM = particulate matter; PM-2.5 = particulate matter with an aerodynamic diameter less than or equal to 2.5 microns; PM-10 = particulate matter with an aerodynamic diameter of less than or equal to 10 microns; SIP = state implementation plan; VOC = volatile organic compound.

[1]Information on any of these emission estimation tools can be obtained from Info CHIEF, (919) 541-5285, info.chief@epamail.epa.gov.

Source: Elsevier Science Inc., adapted from CHIEF.

TABLE 9-4
Locating and Estimating Reports[1]

Pollutant/source category	*EPA and NTIS number*	*Date*
Acrylonitrile	EPA-450/4-84-007a PB84-200609	1984
Arsenic and arsenic compounds	EPA-454/R-98-013	1997
Benzene[2]	EPA-454/R-98-011	1998
1,3-Butadiene[2]	EPA-454/R-96-008 PB97-189880	1997
Cadmium and cadmium compounds[2]	EPA-454/R-93-040 PB94-173093	1993
Carbon tetrachloride[2]	EPA-450/4-84-007b PB84-200625	1984
Chlorobenzenes (revised version)[2]	EPA-454/R-93-044 PB95-111340	1994
Chloroform	EPA-450/4-84-007c PB84-200617	1984
Chromium[2]	EPA-450/4-84-007g PB85-106474	1985
Chromium supplement[2]	EPA-450/2-89-002 PB90-103243	1989
Cyanide compounds[2]	EPA-454/R-93-041 PB94-118379	1993
Dioxins and furans[2]	EPA-454/R-97-003	1997
Epichlorohydrin[2]	EPA-450/4-84-007j PB86-117603	1986
Ethylene dichloride	EPA-450/4-84-007d PB84-239193	1984
Ethylene oxide	EPA-450/4-84-007L PB87-113973	1987
Formaldehyde[2]	EPA-450/4-91-012 PB91-181242	1991
Lead[2]	EPA-454/R-98-006	1998
Manganese	EPA-450/4-84-007h PB86-117587	1986
Mercury and mercury compounds[2]	EPA-454/R-97-012	1997
Methyl chloroform[2]	EPA-454/R-93-045 PB94-214137	1994
Methyl ethyl ketone[2]	EPA-454/R-93-046 PB94-214194	1994
Methylene chloride[2]	EPA-454/R-93-006 PB93-207579	1993
Nickel	EPA-450/4-84-007f PB84-210988	1984
Perchloroethylene and trichloroethylene	EPA-450/2-89-013 PB89-235501	1989
Phosgene	EPA-450/4-84-007i PB86-117595	1986
Polychlorinated biphenyls (PCBs)	EPA-450/4-84-007n PB87-209540	1987
Polycyclic organic matter (POM)[2]	EPA-454/R-98-014	1998
Styrene[2]	EPA-454/R-93-011 PB93-207520	1993
Toluene[2]	EPA-454/R-93-047 PB94-177607	1994
Vinylidene chloride[2]	EPA-450/4-84-007k PB86-117611/	1986
Xylene[2]	EPA-454/R-93-048 PB94-123394	1994

NTIS = National Technical Information Service.

[1]All locating and estimating reports are available from Info CHIEF, (919) 541-5285.

[2]Also available for downloading from the CHIEF Website at http://www.epa.gov/ttn/chief/ap42etc.html#LE.

Source: Adapted from CHIEF.

G. Emission Control Technologies

Three key EPA sources of air emission control technology information (the RBLC, the Clean Air Technology Center [CATC], and the Clean Air Marketplace [CLAM]) are described below.

RBLC

Overview of RACT, BACT, and LAER

To achieve and maintain compliance with the NAAQS, state and local agencies may implement a variety of control and prevention measures, including implementation of "best control" technologies at major existing, new, or modified sources (sometimes certain controls are also required at minor sources). These controls may be implemented in the form of:

- *RACT*—States must require existing sources in certain nonattainment areas to implement RACT to help achieve compliance with the ozone NAAQS.
- *BACT*—Major new sources and modifications to existing sources located in attainment areas are required to obtain a PSD preconstruction permit and implement BACT if the associated emission increase exceeds a specified level. (These sources must apply BACT for each pollutant subject to the CAA.) Furthermore, in nonattainment areas where NAAQS compliance is difficult to achieve, states may require existing sources to implement the more stringent BACT instead of RACT.
- *LAER*—Prior to construction, facilities planning major new sources or modifications to existing sources in nonattainment areas must obtain a nonattainment NSR permit. The permit will require the source to meet a specific LAER imposed in any SIP, even a SIP applying to another locale. In practice, this often means that the source must use emission control measures similar to those specified in southern California, which typically has the most stringent SIP requirements.

Determining the levels of emission control that qualify as RACT, BACT, and LAER can be a difficult task. Specified only in broad, general terms in the NAAQS program provisions, these levels of emission control are based mainly on 1) specific recommendations by EPA, 2) emission control achieved at similar facilities, and 3) other EPA guidance.

EPA has released draft RACT recommendations for several categories of sources. These recommendations are contained in two types of documents: control techniques guidelines (CTGs) and alternative control techniques (ACTs). States can adopt these RACT recommendations, adopt a modified version of the recommendations, or determine RACT on a case-by-case basis. CTGs (and to a lesser extent ACTs) contain a wealth of technical information on typical facility emissions, potential control technologies, and costs. A list of available CTGs and ACTs is presented in Table 3-5 (page 3.6).

Clearinghouse Provides Invaluable Information

The RBLC is an excellent source of RACT, BACT, and LAER information. The clearinghouse, formerly called BLIS, is designed to provide state and local air pollution control agencies with information for making RACT, BACT, and LAER determinations. However, owners/operators can also use the clearinghouse to research candidate technologies likely to qualify for RACT, BACT, or LAER before proposing their selection to the appropriate regulatory agency. (The clearinghouse is also used by regulators to review proposed technology determinations and therefore also provides facilities with a "heads-up" concerning technologies likely to be suggested by regulatory agencies.)

The RBLC contains information distilled from air permits submitted by most of the state and local air pollution control programs in the United States. It contains more than 3,500 RACT, BACT, and LAER determinations from all 50 states and the U.S. territories. Information on emission limits for more than 200 pollutants and 1,000 processes is also included.

RBLC provides a wealth of up-to-date information on facilities for which RACT, BACT, and LAER determinations have been made. It includes:

- *Facility information*—Typically used to search for data on a specific company or to identify determinations that were made in a certain EPA region, state, or locale.
- *Process information*—Ideal for gathering data on determinations made for certain types of processes (e.g., similar sources).
- *Pollutant information*—Determinations for certain pollutants, specific control devices, etc.

The responsibility for determining RACT, BACT, and LAER for a specific facility rests primarily with state and local agencies. However, facility owners/operators often recommend the technology that they believe will satisfy emission control requirements. As always, facility owners/operators who do their homework are likely to come out ahead; that is, keeping up to date with both the developments within an industry and the direction regulators are taking generally pays off.

The RBLC, which is part of the CATC Web site, can be accessed via the Internet at http://mapsweb.rtpnc.epa.gov/RBLCWeb/b102.htm. The site also provides numerous guidance documents (e.g., the *RBLC User Manual*, lists of source codes, SIC codes) that can be downloaded.

CATC

The TTNWeb contains information on air pollution control technologies on the CATC Web site. The CATC provides engineering assistance and technical guidance to state and local air pollution control agencies and, to some extent, industry. The center is maintained by EPA's Air and Engineering Research Laboratory and OAQPS. It is a cooperative effort with the State and Territorial Air Pollution Program Administration (STAPPA) and the Association of Local Air Pollution Control Officials (ALAPCO).

CATC (http://www.epa.gov//ttn/catc) includes access to the RBLC, CATC technical reports that outline industry-specific emissions and/or review control/ management technologies, the *Air Pollution Control Costs Manual*, software, and the CATC Newsletter. (A road map of CATC can be found on page 10.9.) Information can also be obtained from the CATC hotline at (919) 541-0800.

CLAM

Information on the latest developments in air pollution control technologies has been compiled in the CLAM database. The database, created for EPA's Office of Air & Radiation (OAR), provides information on technological innovations, projects, joint ventures, and export opportunities. It contains approximately 1,200 summaries of air pollution control projects compiled from journal articles, press releases, and other public sources. The CLAM database and its user guide can be downloaded from the CLAM Web site (http://www.epa.gov/oar/opar/clam.html). For more information, contact Sabrina Johnson, Office of Policy Analysis and Review, OAR, (202) 260-5285.

H. EPCRA TRI Reporting Resources

EPA's Office of Pollution Prevention and Toxics (OPPT) operates a Web site (http://www.epa.gov/opptintr/tri) that contains valuable toxics release inventory (TRI) information. It includes an overview of the TRI requirements, information on access and use of TRI data, TRI forms and instructions for downloading, reporting requirements, guidance documents, lists of TRI chemicals, and EPA and state TRI contacts.

Other Internet sites that are useful for facilities that must prepare TRIs include:

- *List of lists*—To assist facilities in complying with the chemical reporting requirements under the Emergency Planning and Community Right-to-Know Act (EPCRA), the Comprehensive Environmental Response, Compensation, and Liability Act, and the CAA accidental release prevention program, EPA periodically publishes a "list of lists" that contains all chemicals subject to these reporting requirements. The chemicals are listed alphabetically and by Chemical Abstract Service (CAS) number. An updated list is now available via the Internet at http://www.epa.gov/swercepp/gen-pubs.html (in addition to other EPCRA information). A print copy may be obtained from the RCRA (Resource Conservation and Recovery Act), Superfund & EPCRA Hotline at (800) 424-9346 or (703) 412-9810.
- *Sector-specific guidance*—EPA has issued sector-specific guidance documents to help newly added industries comply with the EPCRA reporting requirements. Guidance documents are available via the Internet (http://www.epa.gov/opptintr/tri/industry.htm) for the following industries: chemical distribution facilities; coal mining; electric-generating facilities; metal mining; petroleum bulk storage facilities; and RCRA hazardous waste treatment, storage, and disposal and solvent recovery facilities.
- *Automated Form R (AFR) software*—To make TRI reporting easier, facilities can use EPA's AFR software to complete electronic Form R submissions. The software (available in DOS and Windows™ versions) can be downloaded from http://www.epa.gov/opptintr/afr. Technical help can be obtained by calling (703) 816-4434 or contacting tris.user.support@epcra.org.

I. Emission Source Databases

The numerous environmental programs regulated by EPA generate an incredible quantity of environmental data on air emission sources. Most of this information is now available online using databases such as the AIRS Web site database. The Plant Information section of the AIRS Web site (http://www.epa.gov/airsweb/sources.htm) contains estimated annual emissions of criteria pollutants, VOCs, and total particulate matter from individual major sources. Reports can be generated for the entire United States, EPA regions, or individual states by specific pollutant. Depending on the report criteria selected, the reports will show: 1) ranking of source emissions, 2) compliance status, 3) facility addresses, 4) number of sources emitting certain pollutants, 5) total pollutant emissions by SIC code, and 6) pollutant emissions reported by year.

The AIRS data can also be accessed via the Envirofacts Warehouse (http://www.epa.gov/enviro). Envirofacts also provides access to 1) compilations of TRI data by facility, year, chemical, and medium of release (air, water, underground injection, land disposal, and offsite), 2) Superfund data on hazardous waste assessment and remediation activities from 1983 to the present, 3) data on hazardous waste generators, transporters, and TSD facilities, 4) National Pollutant Discharge Elimination System water discharge data and permit information, 5) Safe Drinking Water Act information on public water systems and their violations of EPA's safe drinking water regulations, and 6) information on all EPA grant programs. Envirofacts also includes two cross-reference databases: the Facility Index System and the Master Chemical Integrator. In addition, a search engine is available to search for specific information. Finally, maps of environmental information can be generated by choosing from several mapping applications available through EPA's Maps on Demand.

J. Pollution Prevention Information

Enviroene, EPA's pollution prevention information resource, provides a number of valuable resources designed to help industry implement pollution prevention programs. The Web site (http://es.epa.gov) includes:

- *EPA sector notebooks*—EPA sector notebooks for 28 major industrial groups are available for downloading (http://es.epa.gov/oeca/sector/index.html). Each notebook contains an environmental profile, process information, pollution prevention techniques, pollutant release data, regulatory requirements, compliance/enforcement history, innovative programs, and contact names. Industries for which notebooks have been compiled are listed in Table 9-5. In the future, EPA plans to develop notebooks for the following industries: aerospace, agricultural chemicals, agricultural crop production, agricultural stock production, coal mining, local government, and oil and gas exploration and production.

- *Pollution prevention information*—The Technical/Research and Development Information section of the Web site (http://es.epa.gov/techinfo/techinfo.html) includes pollution prevention case studies, fact sheets, searchable bibliographic collections, a list of EPA Pollution Prevention Information Clearinghouse documents, pollution prevention economic information, research project summaries, materials exchange information, and a database of supplemental environmental pollution prevention projects.

- *Solvent alternative data*—Several databases are available to help identify alternatives to hazardous solvents and also provide solvent environmental and safety data.

EPA also publishes the Pollution Prevention Directory (EPA 742-B-94-005, September 1994), a comprehensive guide to pollution prevention programs and contacts. The document can be viewed online at http://www.epa.gov/opptintr/ChemLibPPD.

TABLE 9-5
Industries for Which Industrial Sector Notebooks Are Available[1]

Air transportation
Dry cleaning
Electronics and computers
Fabricated metal products
Federal facilities (draft)
Fossil fuel electric power generation
Ground transportation
Inorganic chemicals
Iron and steel
Lumber and wood products
Metal casting
Metal mining
Motor vehicle assembly
Nonferrous metals
Nonfuel, nonmetal mining
Organic chemicals
Petroleum refining
Pharmaceuticals
Plastic resins and man-made fibers
Printing
Pulp and paper
Rubber and plastics
Shipbuilding and repair
Stone, clay, glass, and concrete
Textiles
Transportation equipment cleaning
Water transportation
Wood furniture and fixtures

[1]Industrial sector notebooks can be accessed from the Enviro-ene Web site at http://es.epa.gov/oeca/sector/index.html.

Source: U.S. EPA.

K. ISO 14000 Contacts

Since the International Organization for Standardization (ISO) issued the first five international environmental management standards in 1996, momentum for initiating the ISO 14000 standards in the United States has been growing steadily. In general, companies that voluntarily adopt the ISO 14000 standards will need to develop an environmental management system (EMS) that demonstrates their commitment to complying with applicable laws/regulations and continually improving their environmental stewardship. Audits must then be performed by certified ISO 14000 auditors to verify compliance.

As a result of the audit requirement, interest in obtaining certification to conduct ISO 14000 EMS audits is skyrocketing. In the United States, auditor training courses must be accredited by the American Society of Quality Registrar Accreditation Board [RAB—(800) 248-1946, rab@rabnet.com]. (RAB also provides auditor certification applications and requirements for a small fee.) At this time, five RAB-certified companies offer ISO 14000 auditor training courses; these are listed in Table 9-6.

Copies of the ISO 14000 standards can be obtained from the American National Standards Institute (ANSI), 11 West 42nd Street, New York, NY 10036, (212) 642-4900. ANSI also maintains a Web site (http://www.ansi.org), as does ISO (http://www.iso.ch).

TABLE 9-6
RAB-Accredited ISO 14000 Auditor Training Courses

Company	*Address and phone number*
Applied Quality Systems	2595 Hamilton Avenue North St. Paul, MN 55113 (612) 633-7902
Excel Partnership, Inc.	75 Glen Road Sandy Hook, CT 06482 (203) 426-3281
International Quality and Environmental Services	41820 West Six Mile Road, Suite 100 Northville, MI 48167 (810) 334-4466
Quality Systems Development	105 Woodside Court Leitchfield, KY 42754 (502) 257-9997
Stat-A-Matrix	One Quality Place Edison, NJ 08820-1059 (908) 548-0600

ANSI = American National Standards Institute; ISO = International Organization for Standardization; RAB = American Society of Quality Registrar Accreditation Board.
Source: ANSI.

L. Risk Assessment Resources

Several risk assessment resources are available via the Internet. Key Web sites include:

- *EPA's Integrated Risk Information System (IRIS)*—IRIS is a database containing information on human health effects that may result from exposure to various substances found in the environment (http://www.epa.gov/ngispgm3/iris). For technical questions, call the EPA Risk Information Hotline at (513) 569-7254.
- *EPA's National Center for Environmental Assessment (NCEA)*—The NCEA Web page (http://www.epa.gov/ncea) provides a number of risk assessment tools and information links (including IRIS, described above). It also offers risk assessment documents, risk guidelines, the Exposure Factors Handbook, and the Integrated Model Evaluation System.
- *Health Effects Notebook for Hazardous Air Pollutants*—This Web site provides fact sheets that describe the human health effects of hazardous air pollutants (http://www.epa.gov/ttn/uatw/hapindex.html). A chemical cross-reference list gives alternative names for the chemicals to allow easier access. In addition, a glossary and summary of health, exposure, and risk assessment terms is provided.
- *National Toxicology Program (NTP)*—The NTP home page offers toxicology testing information, chemical health and safety information, and access to documents and publications (http://ntp-server.niehs.nih.gov).

M. Chemical Databases/Material Safety Data Sheets

Detailed chemical information, such as that provided on material safety data sheets, is now accessible via the Internet. Sites that offer chemical data include:

- *ChemFinder*—This Web site allows access to material safety data sheet information (http://chemfinder.camsoft.com).
- *Hazardous Chemical Database*—Information on 1,991 hazardous chemicals can be gathered from this Web site by keyword (e.g., name, formula, CAS number, Department of Transportation number, RTECS number, EPA number). The database can be accessed via http://ull.chemistry.uakron.edu/erd/sources.html.
- *Office of Prevention, Pesticides, and Toxic Substances (OPPT)*—The OPPT Web site (http://www.epa.gov/chemfact) provides chemical fact sheets for select chemicals. The sheets include information on chemical production and use, environmental fate, and health and environmental effects.

N. Training Courses

Air pollution training courses are offered by a number of sources, including those listed in Table 9-7. For further information on additional training programs, refer to the comprehensive list of training program Web links provided at the STAPPA/ALAPCO Web site (http://www. 4cleanair.org/training.html). The site also features a schedule of upcoming air pollution training courses (http:// www. 4cleanair.org/schedule.html).

Opening up a whole new realm of training possibilities, EPA has begun to offer online air quality courses. An online course on air quality meteorology is available via http://www.shodor.org/metweb/index.html.

O. Enforcement Information

CAA enforcement information can be obtained from many resources, including the Web sites summarized below.

Office of Enforcement and Compliance Assurance

The Office of Enforcement and Compliance Assurance (OECA), which is responsible for ensuring compliance with the CAA and other federal environmental statutes, maintains a Web site that provides CAA compliance information and guidance documents (http://www.epa.gov/oeca). Examples of guidance documents available from the site include EPA's *Audit Policy Interpretive Guidance* (http://es.epa.gov/oeca/apolguid.html) and EPA's *Clean Air Act Policy Compendium* (http://es.epa.gov/oeca/ore/aed/comp/index.html). OECA can also be contacted by calling (202) 564-2440.

Stratospheric Ozone Program Enforcement Actions

This Web site features information on enforcement actions under the stratospheric ozone protection program (http://www.epa.gov/docs/ozone/enforce/enforce.html).

TABLE 9-7
Air Pollution Training Sources

Organization	*Description of courses*	*Phone number*	*Web site address*
Education and Outreach Group (EOG)	EOG, funded by EPA, offers the widest scope of air pollution training in the United States. EPA-sponsored lecture and laboratory courses, using EOG materials, are scheduled at several locations across the country. Self-instructional courses, providing opportunities for individuals to train at home or at their place of employment, are also available.	(919) 515-4265	http://www.epa.gov/oar/oaqps/eog
The Air Pollution Distance Learning Network (APDLN)	APDLN is a state-of-the-art educational satellite broadcasting network made up of more than 100 governmental and university affiliates nationwide. It is a collaborative effort among EPA, state and local air pollution control agencies, and North Carolina State University. The APDLN offers seminars and technical telecourses to state and local employees as well as private sector workers. Many of the courses have been developed by EOG.	(919) 515-4265	http://www.epa.gov/oar/oaqps/eog/apdln.html
Air & Waste Management Association (A&WMA)	A&WMA offers a wide spectrum of air pollution workshops and courses nationwide.	(412) 232-3444	http://www.awma.org/awma/wrkshops/wrkshops.htm
California Air Resources Board (CARB)	The CARB, Compliance Training Section provides practical, usable air compliance training for agency personnel and industry.	Victor Espinosa (916) 322-3976	http://www.arb.ca.gov/cd/ctdmb.htm
Rutgers Air Pollution Training Center	The Rutgers Air Pollution Training Center offers a concentrated air pollution training program for state and local air pollution compliance inspectors. The program is also open to private industry and involves a wide spectrum of courses.	(732) 235-5062	http://www.eohsi.rutgers.edu/cet/airpoll/air.html

Source: Elsevier Science Inc.

P. Environmental Law Resources

The Internet is teeming with terrific environmental law resources. Some of the best Web sites that we came across are described in this section.

EPA Environmental Appeals Board

The full text of all formal decisions issued by EPA's Environmental Appeals Board (EAB) since 1992 are available via EAB's Web site (http://www.epa.gov/eab). The opinions are categorized by program (e.g., CAA penalty appeals, CAA PSD permit appeals). The site also provides general information on EAB and how it makes decisions.

Supreme Court and U.S. Court of Appeals

Decisions handed down by the Supreme Court and U.S. Court of Appeals judges (typically from 1995 onward) are accessible online. A list of these Web sites is presented in Table 9-8. An excellent source of federal court Web links is the Federal Courts Finder, compiled by the Emory University Law Library (http://www.law.emory.edu/FEDCTS). For Web links to state appellate courts (e.g., state supreme courts), refer to the Web links put together by the Cornell University Legal Information Institute (http://supct.law.cornell. edu/supct/directory/overview.html).

State Court Information

Links to state court Web sites are now available via Capital Report's Environmental NewsLink site (http://www.caprep. com/stcourt.htm). Some states provide the full text of opinions online.

Environmental Law Libraries

Table 9-9 presents a list of select environmental law resources on the Internet. These sites provide access to legislation, regulations, court decisions, and much more.

Q. How to Locate and Obtain EPA Publications

EPA publishes a wide array of technical guidance documents. Now, with the advent of online search engines, it's becoming easier and easier to locate and obtain guidance on specific topics. Several main sources of EPA technical guidance information are outlined below.

Publications on the EPA Web Sites

Many EPA Web sites contain technical resource pages that provide direct access to full-text guidance and technical documents. Links to these sites are referenced on the Publications on the EPA Site Web page (http://www.epa.gov/epahome/publications2.htm).

Government Information Locator Service

Although there is an incredible amount of information available via the Internet, ready access is sometimes difficult and time consuming. The Government Information Locator Service (GILS—http://www.epa.gov/gils), an online resource for obtaining EPA information, is extremely useful for locating EPA environmental information. GILS records can be browsed by alphabetical keyword (e.g., air quality criteria, acid rain) by accessing http://www.epa.gov/earth100/browse.

EPA's Online Library System

EPA's Online Library System (OLS) contains references for all of the information resources maintained by the EPA library network. It includes several databases that can be used to locate books, reports, and articles on a variety of topics. The material in OLS is updated every two weeks and can be searched by title, author, subject, keyword, and year of publication. OLS may be accessed via http://www.epa.gov/natlibra/ols.htm.

The most popular database on OLS, the National Catalog, contains the holdings of most of the 28 EPA regional libraries and laboratories. It also features citations and abstracts of EPA reports distributed through the National Technical Information Service (NTIS), regardless of whether they are held by an EPA library.

For technical assistance regarding OLS, contact John Knight at (919) 541-2794 or your nearest EPA library. In addition, a Public Access OLS User Guide can be obtained from EPA's Public Information Center at (202) 260-7751 or (202) 260-2080.

National Technical Information Service

Many of EPA's technical documents are distributed by the NTIS, an agency of the U.S. Department of Commerce. NTIS is the central distribution center for the sale of U.S. and for-

TABLE 9-8
Federal Court Internet Sites

Court	*Web site address*
U.S. Supreme Court	http://supct.law.cornell.edu/supct/directory/united_states.html
U.S. Court of Appeals for the First Circuit	http://www.law.emory.edu/1circuit
U.S. Court of Appeals for the Second Circuit	http://www.TouroLaw.edu/2ndCircuit
U.S. Court of Appeals for the Third Circuit	http://www.law.vill.edu/Fed-Ct/ca03.html
U.S. Court of Appeals for the Fourth Circuit	http://www.law.emory.edu/4circuit
U.S. Court of Appeals for the Fifth Circuit	http://www.ca5.uscourts.gov
U.S. Court of Appeals for the Sixth Circuit	http://www.law.emory.edu/6circuit
U.S. Court of Appeals for the Seventh Circuit	http://www.kentlaw.edu/7circuit
U.S. Court of Appeals for the Eighth Circuit	http://www.wulaw.wustl.edu/8th.cir
U.S. Court of Appeals for the Ninth Circuit	http://www.findlaw.com/casecode/courts/9th.html
U.S. Court of Appeals for the Tenth Circuit	http://www.law.emory.edu/10circuit
U.S. Court of Appeals for the Eleventh Circuit	http://www.law.emory.edu/11circuit
U.S. Court of Appeals for the District of Columbia Circuit	http://www.ll.georgetown.edu/Fed-Ct/cadc.html

Source: Elsevier Science Inc.

TABLE 9-9
Environmental Law Libraries on the Internet

Name	*Description of available resources*	*Web site address*
Cornell University Legal Information Institute	*U.S. Code*, *CFR*, recent U.S. Supreme Court environmental decisions, state environmental regulations, and many other environmental law Web sites	http://www.law.cornell.edu/topics/environmental.html
Environmental Law World Wide Web Site	Recent environmental opinions from federal courts, new and proposed EPA rules, EPA guidance documents, and new and proposed legislation	http://www.webcom.com/~staber/rectinfo.html
LEXIS-NEXIS Environmental Law Library	Wide array of environmental law materials available for a subscription fee, including federal and state case law, federal and state statutes, *FR*, legislative histories, *Congressional Record*, environmental law reviews, publications from The Bureau of National Affairs, Inc., and more than 70 leading industry publications (including *The Air Pollution Consultant*)	http://www.lexis-nexis.com/lncc/sources/libcont/envirn.html
Pace Virtual Environmental Law Library	Links to Internet legal sources for U.S. and international environmental law	http://willy.law.pace.edu/env/vell6.html
U.S. House of Representatives Internet Law Library	Federal and state laws, laws of other nations, treaties and international law, attorney and legal profession directories, law book reviews, and law school library catalogs	http://law.house.gov
Washburn University School of Law's Web Site	Comprehensive resource for information on federal law, international law, law journals, law library catalogs, and state laws/agencies; search engine for locating case law	http://lawlib.wuacc.edu/washlaw/washlaw.intro.html

CFR = *Code of Federal Regulations*; *FR* = *Federal Register*.

Source: Elsevier Science Inc.

eign government-sponsored research, development, engineering, and business reports.

Approximately 70,000 technical reports are added annually to the NTIS database, which contains almost 3 million titles (including environmental and other publications). The NTIS bibliographic database can be searched online via http://www.ntis.gov/search.htm.

NTIS assigns unique identification numbers to every document that it distributes. For example, many EPA documents are assigned an NTIS number that begins with "PB." The NTIS number should be used when placing an order; however, documents can sometimes be located using the EPA number. Orders can be placed by contacting NTIS at 5285 Port Royal Road, Springfield, VA 22161, (800) 553-6847 or (703) 605-6000.

National Center for Environmental Publications and Information

The National Center for Environmental Publications and Information (NCEPI) is a central repository for all EPA documents; more than 5,500 titles are available. Using NCEPI's Web server (http://www.epa.gov/ncepihom/index.html), you can browse and search NCEPI's National Publications Catalog. EPA publications can then be ordered either online or by calling (800) 490-9198.

National Environmental Publications Information

Using EPA's largest electronic publications Web site, you can search, view, and print from a collection of more than 6,000 technical and public information publications. The National Environmental Publications Information database is accessible via http://www.epa.gov/cincl.

Government Printing Office

Key regulatory compliance references, such as the *FR* and the *CFR*, are available from the GPO. The GPO can be contacted by calling (202) 512-1800 or accessing the GPO Web site (http://www.access.gpo.gov/su_docs/index.html). GPO also operates 24 U.S. government bookstores throughout the country. For the bookstore nearest you, call GPO or access http://www.access.gpo.gov/su_docs/ sale/abkst001.html.

R. EPA Contacts

EPA's OAR is the ultimate source of information on CAA issues. OAR, which is responsible for implementing the CAA, is comprised of four groups: OAQPS, the Office of Atmospheric Programs (OAP), the Office of Mobile Sources, and the Office of Radiation and Indoor Air. A list of select OAR contacts, including divisions within OAQPS and OAP, is presented in Table 9-10. EPA regional contacts are provided in Table 9-11 (page 9.23).

Further EPA contact information may be obtained at the following Web sites:

- *EPA Web page*—Links to EPA offices, labs, and regions are accessible via http://www.epa.gov/epahome/Offices.html. Other resources (e.g., libraries, dockets, clearinghouses, hotlines, models) can be accessed at http://www.epa.gov/epahome/finding.html.
- *OAR Information Resources Document*—The *OAR Information Resources Document* provides an exhaustive list of OAR contacts and resources. A .pdf version of the publication is available via http://www.epa.gov/oar/oarfacts.html.
- *EPA people locator*—The EPA people locator database is a valuable resource for finding specific individuals within EPA (http://www.epa.gov:6706/natlocdcd/owa/locator.home). You simply type in the first (optional) and last name of the person you are seeking, and the database will locate information about them, including their phone number, EPA address, and e-mail address.

TABLE 9-10
Directory of EPA OAR[1]

Office	Responsibilities	Phone number	Web site address
Office of Air Quality Planning and Standards (OAQPS)	• Develops regulations • Compiles and reviews air pollution data • Assists states and local agencies with monitoring and air pollution controls • Makes air pollution information available to the public	(919) 541-5616	http://www.epa.gov/oar/oaqps
Air Quality Strategies and Standards Division	• Develops national and geographically focused strategies and programs for air quality management based on assessments of health and ecological effects, exposure and risk, and economic impacts and benefits	(919) 541-5551	http://www.epa.gov/oar/oaqps/org/aqssd
Emissions, Monitoring, and Analysis Division	• Develops and distributes guidelines for air quality models and monitoring • Establishes air quality indicators • Analyzes air pollution trends • Develops emission factors • Provides technical guidance on emission inventories	(919) 541-5536	http://www.epa.gov/oar/oaqps/org/emad
Emission Standards Division	• Establishes emission standards for hazardous air pollutants and criteria air pollutants, new source performance standards, control techniques guidelines, alternative control techniques documents, and technical guidance • Provides technical assistance on effective control technologies	(919) 541-5572	http://www.epa.gov/oar/oaqps/org/esd
Information Transfer and Program Integration Division	• Integrates operating permit program with air programs • Transfers air pollution control information • Develops and conducts training courses	(919) 541-5604	http://www.epa.gov/oar/oaqps/org/itpid

TABLE 9-10
Directory of EPA's Office of Air and Radiation Offices[1]—Continued

Office	*Responsibilities*	*Phone number*	*Web site address*
Office of Atmospheric Programs	Implements the acid rain program and the stratospheric ozone program (which consists of the ozone layer protection program and the atmospheric pollution prevention program)	(202) 564-9140	http://www.epa.gov/oar/oap.html
Acid Rain Program	Requires emission reductions of sulfur dioxide and nitrogen oxides from electric utilities	(202) 564-9150	http://www.epa.gov/acidrain/ardhome.html
Ozone Layer Protection Program	Regulates the phase out of of ozone-depleting substances such as chlorofluorocarbons, methyl chloroform, and methyl bromide	(202) 564-9101	http://www.epa.gov/ozone
Atmospheric Pollution Prevention Program	Focuses on reducing greenhouse gas emissions through implementation of voluntary energy efficiency programs	(202) 564-9190	http://www.epa.gov/appdstar/appd
Office of Mobile Sources	Regulates emissions from motor vehicles, fuels, and engines	(202) 260-7645	http://www.epa.gov/OMSWWW/omshome.htm
Office of Radiation and Indoor Air	Protects public health and indoor air pollution	(703) 356-4020	http://www.epa.gov/iaq

OAR = Office of Air & Radiation.

[1]OAR, which is responsible for implementation of the CAA, can be contacted by calling (202) 260-7400 or by accessing the OAR web site at http://www.epa.gov/oar.

Source: Elsevier Science Inc.

TABLE 9-11
Regional EPA Contacts

Region	*Address*	*States covered*
Region I	John F. Kennedy Building Boston, MA 02203 (888) 372-7341 http://www.epa.gov/region1	Connecticut, Maine, Massachusetts, New Hampshire, Rhode Island, and Vermont
Region II	290 Broadway New York, NY 10007 (212) 637-3000 http://www.epa.gov/region2	New Jersey, New York, Puerto Rico, and the U.S. Virgin Islands
Region III	1650 Arch Street Philadelphia, PA 19103-2029 (215) 814-5000 http://www.epa.gov/region3	Delaware, District of Columbia, Maryland, Pennsylvania, Virginia, and West Virginia
Region IV	Atlanta Federal Center 61 Forsyth Street, SW Atlanta, GA 30303-3104 (404) 562-9900 http://www.epa.gov/region4	Alabama, Florida, Georgia, Kentucky, Mississippi, North Carolina, South Carolina, and Tennessee
Region V	77 West Jackson, Blvd. Chicago, IL 60604 (312) 363-2000 http://www.epa.gov/region5/	Illinois, Indiana, Michigan, Minnesota, Ohio, and Wisconsin
Region VI	1445 Ross Avenue, Suite 1200 Dallas, TX 75202 (214) 665-2200 http://www.epa.gov/earth1r6/index.htm	Arkansas, Louisiana, New Mexico, Oklahoma, and Texas
Region VII	726 Minnesota Ave. Kansas City, KS 66101 (913) 551-7003 http://www.epa.gov/region7/	Iowa, Kansas, Missouri, and Nebraska
Region VIII	999 18th Street, Suite 500 Denver, CO 80202-2466 (303) 312-6312 http://www.epa.gov/unix0008	Colorado, Montana, North Dakota, South Dakota, Utah, Wyoming, and 27 Tribal Nations
Region IX	75 Hawthorne Street San Francisco, CA 94105 (415) 744-1500 http://www.epa.gov/region9	Arizona, California, Hawaii, Nevada, the Pacific Islands subject to U.S. law, and approximately 140 Tribal Nations
Region X	1200 6th Avenue Seattle, WA 98101 (206) 553-1200 http://www.epa.gov/docs/region10/	Alaska, Idaho, Oregon, and Washington

Source: Elsevier Science Inc.

S. State, Local, and Regional Contacts

State and Local Agency Contacts

Table 9-12 is a list of state and local air pollution agency contacts, including addresses, phone numbers, and Web site addresses. For current information on state Web sites, we suggest checking the Air Pollution Control Resources site (http://members.aol.com/CleenAir/apc/state.htm). This resource not only includes Web links, but also describes the status and content of each state/local agency Web site. Other helpful state information Web sites include:

- The State Web Locator operated by the Villanova Center for Information Law and Policy (http://www.law.vill.edu/State-Agency/index.html);
- A state law search engine operated by Washburn University School of Law (http://lawlib.wuacc.edu/washlaw/uslaw/statelaw.html);
- StateSearch sponsored by the National Association of State Information Resource Executives (http://www.nasire.org/ss/index.html);
- The STAPPA/ALAPCO Web site (http://www.4cleanair.org/states.html); and
- Links to state publications and fact sheets designed to help businesses with environmental compliance and pollution prevention issues provided by the Small Business home page. The fact sheets are listed either by state (http://www.icubed.com/epa_sbo/html/programs_statepub.html) or industry sector (http://www.icubed.com/epa_sbo/html/indsector.html).

Regional Air Pollution Control Initiatives

Summaries and Web links for a number of regional air pollution control initiatives (e.g., the Great Lakes Commission) are provided at http://members.aol.com/CleenAir/apc/fed2.htm#regional. The Ozone Transport Assessment Group, which was formed to study the effect of ozone transport on nonattainment areas, has its own Web site under the TTNWeb (http://www.epa.gov/otag).

TABLE 9-12
State and Local Air Pollution Control Agency Contacts

Alabama

Alabama Department of Environmental Management
Air Division
1751 Congressman W.L. Dickinson Drive
Montgomery, AL 36109-2608
http://www.adem.state.al.us/
(334) 271-7861

Local agencies:

Environmental Health Services
Jefferson County Department of Health
P.O. Box 2648
Birmingham, AL 35202-2648
(205) 930-1292

Natural Resources and Environmental Management Division
City of Huntsville
305 Church Street
Huntsville, AL 35801
(256) 535-4206
http://www.ci.huntsville.al.us/NatRes

Alaska

Alaska Department of Environmental Conservation
Division of Air and Water Quality
Air Quality Improvement Section
410 Willoughby Avenue, Suite 105
Juneau, AK 99801-1795
(907) 465-5300
http://www.state.ak.us/local/akpages/ENV.CONSERV/dawq/dec_dawq.htm

Arizona

Arizona Department of Environmental Quality
Air Quality Division
3033 N. Central Avenue
Phoenix, AZ 85012
(602) 207-2308
http://www.adeq.state.az.us/air/index.htm

TABLE 9-12
State and Local Air Pollution Control Agency Contacts—Continued

Arizona—Continued

Local agencies:

Maricopa County Environmental Services
01001 N. Central Avenue
Phoenix, AZ 85004
(602) 506-6700

Pima County Department of Environmental Quality
130 W. Congress
Tucson, AZ 85701
(520) 740-3340
http://www.deq.co.pima.az.us

Pinal County Air Quality Control District
P.O. Box 987
Florence, AZ 85232
(520) 868-6762

Arkansas

Arkansas Department of Pollution Control and Ecology
(will become Arkansas Department of Environmental Quality on 3/31/99)
Air Division
8001 National Drive
Little Rock, AR 72209
(501) 682-0730

California

California Air Resources Board
2020 L Street
Sacramento, CA 95814
(800) 272-4572 (in CA)
(916) 323-3336 (outside CA)
http://www.arb.ca.gov/homepage.htm

Local agencies:

Amador County APCD
500 Argonaut Lane
Jackson, CA 95642-2310
(209) 223-6406

Antelope Valley APCD
(northeast portion of LA County)
43301 Division Street, Suite 206
Lancaster, CA 93539-4409
(805) 723-8070
http://www.mdaqmd.ca.gov

Bay Area AQMD
(Alameda, Contra Costa, Marin, Napa, San Francisco, San Mateo, Santa Clara, portions of Solano, and Sonoma Counties)
939 Ellis Street
San Francisco, CA 94109-7714
(415) 771-6000
http://www.baaqmd.gov

Butte County AQMD
2525 Dominic Drive, Suite J
Chico, CA 95928-7184
(530) 891-2882
http://www.dcn.davis.ca.us/~bluesky

Calaveras County APCD
891 Mountain Ranch Road
San Andreas, CA 95249-9709
(209) 754-6504

Colusa County APCD
100 Sunrise Blvd, Suite F
Colusa, CA 95932-3246
(530) 458-0590
http://www.den.davis.ca.us/~bluesky

El Dorado County APCD
2850 Fairlane Court, Bldg. C
Placerville, CA 95667-4100
(530) 621-6662

Feather River AQMD
(Yuba and Sutter Counties)
938 14th Street
Marysville, CA 95901-4149
(530) 634-7659
http://www.dcn.davis.ca.us/~bluesky

TABLE 9-12
State and Local Air Pollution Control Agency Contacts—Continued

California—Continued

Local agencies:—Continued

Glenn County APCD
P.O. Box 351
720 North Colusa Street
Willows, CA 95988-0351
(530) 934-6500

Great Basin Unified APCD
(Mono, Inyo, and Alpine Counties)
157 Short Street, Suite 6
Bishop, CA 93514-3537
(760) 872-8211

Imperial County APCD
150 South 9th Street
El Centro, CA 92243-2801
(760) 339-4314

Kern County APCD
(eastern portion of Kern County)
2700 M Street, Suite 302
Bakersfield, CA 93301-2370
(805) 862-5250

Lake County AQMD
883 Lakeport Blvd
Lakeport, CA 95453-5405
(707) 263-7000

Lassen County APCD
175 Russle Avenue
Susanville, CA 96130-4215
(530) 251-8110

Mariposa County APCD
P.O. Box 2039
5101 Jones Street
Mariposa, CA 95338-2039
(209) 966-5151

Mendocino County AQMD
306 East Gobbi Street
Ukiah, CA 95482-5511
(707) 463-4354

Modoc County APCD
202 West 4th Street
Alturas, CA 96101-3915
(530) 233-6419

Mojave Desert AQMD
(northern portion of San Bernardino County, eastern portion of Riverside County)
15428 Civic Drive, Suite 200
Victorville, CA 92392-2383
(760) 245-1661
http://www.mdaqmd.ca.gov

Monterey Bay Unified APCD
(Monterey, San Benito, and Santa Cruz Counties)
24580 Silver Cloud Court
Monterey, CA 93940-6536
(831) 647-9411

North Coast Unified AQMD
(Humboldt, Del Norte, and Trinity Counties)
2300 Myrtle Avenue
Eureka, CA 95501-3327
(707) 443-3093

Northern Sierra AQMD
(Plumas, Sierra, and Nevada Counties)
P.O. Box 2509
200 Litton Drive, Suite 320
Grass Valley, CA 95945-2509
(530) 274-9360
http://www.nccn.net/~nsaqmd

Northern Sonoma County APCD
(northern portion of Sonoma County)
150 Matheson Street
Healdsburg, CA 95448-4908
(707) 433-5911

Placer County APCD
11464 B Avenue—DeWitt Center
Auburn, CA 95603-2603
(530) 889-7130
http://www.dcn.davis.ca.us/~bluesky

TABLE 9-12
State and Local Air Pollution Control Agency Contacts—Continued

California—Continued

Local agencies:—Continued

Sacramento Metropolitan AQMD
8411 Jackson Road
Sacramento, CA 95826-3904
(916) 386-6650
http://www.airquality.org *or*
http://www.sparetheair.com

San Diego County APCD
9150 Chesapeake Drive
San Diego, CA 92123-1096
(619) 694-3300
http://www.sdapcd.co.san-diego.ca.us

San Joaquin Valley Unified APCD
(Fresno, Kings, Kern [part], Madera, Merced, San Joaquin, Stanislaus, and Tulare Counties)
1999 Tuolumne Street, Suite 200
Fresno, CA 93721-1638
(209) 497-1000

San Luis Obispo County APCD
3433 Roberto Court
San Luis Obispo, CA 93401-7126
(805) 781-5912
http://www.sloapcd.dst.ca.us

Santa Barbara County APCD
26 Castilian Drive, Suite B-23
Goleta, CA 93117-3027
(805) 961-8800
http://www.santa-barbara.ca.us/~apcd

Shasta County AQMD
1855 Placer Street, Suite 101
Redding, CA 96001-1759
(530) 225-5674
http://www.dcn.davis.ca.us/~bluesky

Siskiyou County APCD
525 South Foothill Drive
Yreka, CA 96097-3036
(530) 841-4029

South Coast AQMD
(Los Angeles [except Antelope APCD], Orange, western portion of San Bernardino, and western portion of Riverside Counties)
21865 E. Copley Drive
Diamond Bar, CA 91765-4182
(909) 396-2000
http://www.aqmd.gov

Tehama County APCD
P.O. Box 38
1750 Walnut Street
Red Bluff, CA 96080-0038
(530) 527-3717

Tuolumne County APCD
22365 Airport
Columbia, CA 95310
(209) 533-5693

Ventura County APCD
669 County Square Drive, 2nd Floor
Ventura, CA 93003-5417
(805) 645-1400

Yolo-Solano AQMD
(Yolo County and eastern part of Solano County)
1947 Galileo Court, Suite 103
Davis, CA 95616-4882
(530) 757-3650
http://www.dcn.davis.ca.us/~ysaqmd

Colorado

Colorado Department of Public Health and Environment
APCD-SS-B1
4300 S. Cherry Creek Drive
Denver, CO 80246-1530
(303) 692-3100

Local agency:

Regional Air Quality Council
(Denver metro area)
1445 Market Street #260
Denver, CO 80202
(303) 629-5450
http://www.raqc.org

TABLE 9-12
State and Local Air Pollution Control Agency Contacts—Continued

Connecticut

Connecticut Department of Environmental Protection
Bureau of Air Management
79 Elm Street
Hartford, CT 06106-5127
(860) 424-3026
http://dep.state.ct.us/air/airhome.htm

Delaware

Delaware Department of Natural Resources and Environmental Control
Air Quality Management Section
715 Grantham Lane
New Castle, DE 19720
(302) 323-4542
http://www.dnrec.state.de.us/air/aqm_page/aqm_nets.htm

District of Columbia

Air Resources Management Division
Environmental Regulation Administration
2100 Martin Luther King Jr. Avenue SE, Suite 404
Washington, DC 20020
(202) 645-6093

Florida

Florida Department of Environmental Protection
Division of Air Resources Management
2600 Blair Stone Road, Mail Station 5500
Tallahassee, FL 32399-2400
(850) 488-0114
http://www.dep.state.fl.us/air/index.htm

Georgia

Georgia Department of Natural Resources
Air Protection Branch
4244 International Parkway, Suite 120
Atlanta, GA 30354
(404) 363-7000
http://ganet.ganet.org/dur/environ/branches

Hawaii

Hawaii Department of Health
Clean Air Branch
919 AlaMoana Blvd, Suite 203
Honolulu, HI 96814
(808) 586-4200
http://www.hawaii.gov/health/ei/emca/eiemca00.htm

Idaho

Idaho Division of Environmental Quality Permits and Enforcement
1410 N. Hilton
Boise, ID 83706
(208) 373-0502
http://www2.state.id.us/deq/air.htm

Illinois

Illinois Environmental Protection Agency
Bureau of Air
P.O. Box 19276
1021 N. Grand Avenue East
Springfield, IL 62794-9276
(217) 785-4140
http://www.epa.state.il.us/about/org/bureau-of-air.html

Indiana

Indiana Department of Environmental Management
Office of Air Management
100 North Senate
P.O. Box 6015
Indianapolis, IN 46206-6015
(317) 233-0178
http://www.state.in.us/idem/oam/index.html

TABLE 9-12
State and Local Air Pollution Control Agency Contacts—Continued

Iowa

Iowa Department of Natural Resources
Air Quality Bureau
7900 Hickman Road, Suite 1
Urbandale, IA 50322
(515) 281-5100
http://www.state.ia.us/government/dnr/organiza/epd/airq/aqbur.htm

Local agencies:

Linn County Health Department
Air Quality Division
501 13th Street NW
Cedar Rapids, IA 52405-3700
(319) 398-3551
http://www.air.linn.ia.us

Polk County Air Quality Division
1595 NE 58th Avenue
Des Moines, IA 50313
(515) 286-3351
http://www.co.polk.ia.us/airquality/welcome.htm

Kansas

Kansas Department of Health and Environment
Bureau of Air and Radiation
Forbes Field, Building 283
Topeka, KS 66620-0001
(785) 296-5193
http://www.ink.org/public/kdhe/bar/

Local agency:

United government of Wyandotte County
Department of Air Quality
619 Ann Avenue
Kansas City, KS 66101
(913) 573-6700
http://www.toto.net/daq

Kentucky

Kentucky Department for Environmental Protection
Division for Air Quality
803 Schenkel Lane
Frankfort, KY 40601-1403
(502) 573-3382
http://www.state.ky.us/agencies/nrepc/dep/daq/daqhome.html

Local agency:

Jefferson County Air Pollution Control District
850 Barret Avenue
Louisville, KY 40204-1745
(502) 574-6000
http://www.apcd.org

Louisiana

Louisiana Department of Environmental Quality
Office of Air Quality and Radiation Protection
P.O. Box 82135
Baton Rouge, LA 70884-2135
(225) 765-0219
http://www.deq.state.la.us/oarp/oarp.htm

Maine

Maine Department of Environmental Protection
Bureau of Air Quality
17 State House Station
Augusta, ME 04333-0017
(207) 287-2437
http://www.state.me.us/dep/air/home.htm

Maryland

Maryland Department of the Environment
Air and Radiation Management Administration
2500 Broening Highway
Baltimore, MD 21224
(410) 631-3225
http://www.mde.state.md.us/arma/airqulty/airqual.html

TABLE 9-12
State and Local Air Pollution Control Agency Contacts—Continued

Massachusetts

Massachusetts Department of Environmental Protection
One Winter Street
Boston, MA 02108
(617) 292-5500
http://www.magnet.state.ma.us/dep/

Michigan

Michigan Department of Environmental Quality
Air Quality Division
Holister Building, 4th Floor
P.O. Box 30260
106 West Allegan
Lansing, MI 48909-7760
(517) 373-7023
http://www.deq.state.mi.us/aqd/

Minnesota

Minnesota Pollution Control Agency
Division of Air Quality
520 Lafayette Road
St. Paul, MN 55155-4194
(651) 296-6300
http://www.pca.state.mn.us/air/index.html

Mississippi

Mississippi Department of Environmental Quality
Office of Pollution Control
P.O. Box 10385
Jackson, MS 39289-0385
(601) 961-5171
http://www.deq.state.ms.us/domino/pcweb.nsf

Missouri

Missouri Department of Natural Resources
Air Pollution Control Program
P.O. Box 176
Jefferson City, MO 65102
(800) 334-6946
http://www.dnr.state.mo.us/deq/apcp/homeapcp.htm

Local agencies:

City of St. Louis
Division of Air Pollution Control
1220 Carr Lane Avenue
St. Louis, MO 63104-1010
(314) 664-7877
http://stlouis.missouri.org/citygov/airpollution/index.htm

Montana

Montana Department of Environmental Quality
Air & Waste Management Bureau
P.O. Box 200901
1520 East Sixth Avenue
Helena, MT 59620-0901
(406) 444-2544
http://www.deq.state.mt.us/pcd/awm/index.htm

Nebraska

Nebraska Department of Environmental Quality
Air Quality Section
P.O. Box 98922
Lincoln, NE 68509-8922
(402) 471-2186
http://www.deq.state.ne.us/contents.nsf

Local agencies:

Lincoln-Lancaster County Health Department
Environmental Health Division/Air Programs
3140 N Street
Lincoln, NE 68510
(402) 441-8021
http://interlinc.ci.lincoln.ne.us/interlinc/city/health/environ/pollu/index.htm

Douglas County Health Department
Air Quality Control Division
1819 Farnam, Room 401
Omaha, NE 68183
(402) 444-7485

TABLE 9-12
State and Local Air Pollution Control Agency Contacts—Continued

Nevada

Nevada Division of Environmental Protection
Bureau of Air Quality
333 West Nye Lane
Carson City, NV 89706
(702) 687-4670

Local agencies:

Clark County Health District
Air Pollution Control Division
P.O. Box 3902
Las Vegas, NV 89127
(702) 383-1276

Washoe County District Health Department
Air Quality Management Division
P.O. Box 11130
Reno, NV 89520-0027
(702) 784-7200

New Hampshire

New Hampshire Department of Environmental Services
Air Resources Division
64 N. Main Street
Concord, NH 03302-2033
(603) 271-1370
http://www.state.nh.us/des/ard/homepage.htm

New Jersey

New Jersey Department of Environmental Protection
Air Quality Permitting Program
P.O. Box 027
Trenton, NJ 08625-0027
(609) 633-2829
http://www.state.nj.us/dep/aqpp/

New Mexico

New Mexico Environment Department
Air Quality Bureau
1190 St. Francis Drive
Santa Fe, NM 87502
(505) 827-0031
http://www.nmenv.state.nm.us/

Local agency:

City of Albuquerque, Environmental Health Department
Air Pollution Control Division
P.O. Box 1293
Albuquerque, NM 87103
(505) 768-2600

New York

New York Department of Environmental Conservation
Division of Air Resources
50 Wolf Road
Albany, NY 12233-3250
(518) 457-7230
http://www.dec.state.ny.us/website/dar/index.html

North Carolina

North Carolina Department of Environment
and Natural Resources
Division of Air Quality
P.O. Box 29580
Raleigh, NC 27626-0580
(919) 733-3340
http://www.ehnr.state.nc.us/ehnr/aq/

Local agencies:

Forsythe County Environmental Affairs
537 N. Spruce Street
Winston–Salem, NC 27101-1362
(336) 727-8060
http:www.co.forsyth.nc.us/envaffairs/envaffairs.htm

TABLE 9-12
State and Local Air Pollution Control Agency Contacts—Continued

North Carolina

Local agencies:—Continued

Mecklenburg County Department of Environmental Protection
700 N. Tryon Street, Suite 205
Charlotte, NC 28202
(704) 336-5500
http://www.charmeck.nc.us/coenv/home.htm

Western North Carolina Regional Air Pollution Control Agency
49 Mt. Carmel Road
Asheville, NC 28806
(704) 255-5655

North Dakota

North Dakota Department of Health
Division of Environmental Engineering
1200 Missouri Avenue, Room 304
Bismarck, ND 58506-5520
(701) 328-5188
http://www.health.state.nd.us/ndhd/environ/ee/index.htm

Ohio

Ohio Environmental Protection Agency
Division of Air Pollution Control
P.O. Box 1049
Columbus, OH 43216-1049
(614) 644-2270
http://www.epa.ohio.gov/dapc/dapcmain.html

Local agencies:

Air Pollution Group
Portsmouth City Health Dept.
740 Second Street
Portsmouth, OH 45662

Akron Air Pollution Control
146 South High Street, Room 904
Akron, OH 44308
(330) 375-2480
http://members.tripod.com/~araqmd/main/

Bureau of Engineering Services
Division of Air Pollution Control
1925 St. Clair
Cleveland, OH 44114
(216) 664-2324

Department of Environmental Services
Air Quality Program
1632 Central Parkway
Cincinnati, OH 45210
(513) 651-9437

Division of Air Pollution Control
Canton City Health Department
420 Market Avenue, N.
Canton, OH 44702-1544
(330) 489-3385 or 489-3231
http://198.234.65.72/cchd/serv03.htm

Regional Air Pollution Control Agency
County Administration Building
451 W. Third Street
Dayton, OH 45422
(937) 225-4435
http://laa.co.montgomery.oh.us/rapca.htm

Toledo Environmental Control
348 South Erie
Toledo, OH 43602
(419) 936-3015

Oklahoma

Oklahoma Department of Environmental Quality
Air Quality Division
P.O. Box 1677
Oklahoma City, OK 73101-1677
(405) 702-4100
http://www.deq.state.ok.us/air1/air.html

TABLE 9-12
State and Local Air Pollution Control Agency Contacts—Continued

Oregon

Oregon Department of Environmental Quality
Air Quality Division
811 SW Sixth Avenue
Portland, OR 97204-1390
(503) 229-5696
http://www.deq.state.or.us/aq/aq.htm

Local agency:

Lane Regional Air Pollution Authority
1010 Main Street
Springfield, OR 97477
(541) 726-1930

Pennsylvania

Pennsylvania Department of Environmental Protection
Bureau of Air Quality
P.O. Box 8468
Harrisburg, PA 17105-8468
(717) 787-9702
http://www.dep.state.pa.us/dep/deputate/airwaste/aq/default.htm

Local agencies:

Philadelphia Air Management Services
321 University Avenue
Spelman Building
Philadelphia, PA 19104
(215) 823-7584

Allegheny County Health Department
Division of Air Pollution Control
301 39th Street, Bldg. #7
Pittsburgh, PA 15201-1891
(412) 578-8111

Rhode Island

Rhode Island Department of Environmental Management
Division of Air Resources
235 Promenade Street
Providence, RI 02908-5767
(401) 222-2808
http://www.state.ri.us/dem

South Carolina

South Carolina Department of Health and Environmental Control
Bureau of Air Quality
2600 Bull Street
Columbia, SC 29201
(803) 734-4750
http://www.state.sc.us/dhec/baq_home.htm

South Dakota

South Dakota Department of Environment and Natural Resources
Air and Surface Water Program
523 East Capital
Pierre, SD 57501
(605) 773-7171
http://www.state.sd.us/state/executive/denr/des/airquality/airprogr.htm

Tennessee

Tennessee Department of Environment and Conservation
Division of Air Pollution Control
9th Floor, L&C Annex
401 Church Street
Nashville, TN 37243-1531
(615) 532-0554
http://www.state.tn.us/environment/apc/index.html

Local agencies:

Chattanooga-Hamilton County
Air Pollution Control Bureau
3511 Rossville Blvd.
Chattanooga, TN 37407
(423) 867-4321

Knox County Air Pollution Control
City County Building, Room 339
400 Main Avenue
Knoxville, TN 37902-2405
(423) 215-2488

TABLE 9-12
State and Local Air Pollution Control Agency Contacts—Continued

Tennessee

Local agencies:—Continued

Memphis/Shelby County Health Department
Pollution Control Section
4th Floor
814 Jefferson Avenue
Memphis, TN 38105
(901) 544-7775

Metro Health Department
Air Pollution Control Division
311 23rd Avenue North
Nashville, TN 37203
(615) 340-5653

Texas

Texas Natural Resources Conservation Commission
Office of Air Quality
P.O. Box 13087, MC-161
Austin, TX 78711-3087
(512) 239-5440
http://www.tnrcc.texas.gov/air/index.html

Local agency:

Fort Worth Environmental Management Department
1000 Throckmorton Street
Fort Worth, TX 76102
(817) 871-8079
http://ci.fort-worth.tx.us/fortworth/dem/index.htm

Utah

Utah Department of Environmental Quality
Division of Air Quality
P.O. Box 144820
150 North 1950 West
Salt Lake City, UT 84114-4820
(801) 536-4000
http://www.eq.state.ut.us/eqair/aqhome.htm

Vermont

Vermont Agency of Natural Resources
Air Pollution Control Division
103 South Main Street, Building #3 South
Waterbury, VT 05671-0402
(802) 241-3840
http://www.anr.state.vt.us/dec/air/apcdhome.htm

Virginia

Virginia Department of Environmental Quality
Air Programs
P.O. Box 10009
Richmond, VA 23240
(804) 698-4000
http://www.deq.state.va.us/envprog/airqual.html

Washington

Washington State Department of Ecology
Air Quality Program
P.O. Box 47600
Olympia, WA 98504-7600
(360) 407-6800
http://www.wa.gov/ecology/air/airhome.html

Local agencies:

Benton County Clean Air Authority
650 George Washington Way
Richland, WA 99352-4290
(509) 943-3396
http://www.cbvcp.com/bccaa

Northwest Air Pollution Authority
1600 S. 2nd Street
Mt. Vernon, WA 98273-5202
(360) 428-1617
http://www.pacificrim.net/~nwapa

Olympic Air Pollution Control Authority
909 Sleater-Kinney Road SE, Suite 1
Lacey, WA 98503
(360) 438-8768
http://www.wln.com/~oapca

TABLE 9-12
State and Local Air Pollution Control Agency Contacts—Continued

Washington

Local agencies:—Continued

Puget Sound Air Pollution Control Authority
110 Union Street, Suite 500
Seattle, WA 98101
(206) 343-8800
http://www.psapca.org/

Southwest Air Pollution Control Authority
1308 NE 134th Street
Vancouver, WA 98685-2747
(360) 574-3058
http://www.swapca.org/

Spokane County Air Pollution Control Authority
1101 West College Avenue, Suite 403
Spokane, WA 99201
(509) 477-4727
http://www.scapca.org/

Yakima County Clean Air Authority
6 South 2nd Street, Room 1016
Yakima, WA 98901
(509) 574-1410
http://www.wa.gov/ecology/air/yrcaa.html

West Virginia

West Virginia Division of Environmental Protection
Office of Air Quality
1558 Washington Street East
Charleston, WV 25311
(304) 558-4022
http://wvnvm.wvnet.edu/~jmorgan/index.htm

Wisconsin

Wisconsin Dept. of Natural Resources
Bureau of Air Management
Gef 2 Building, 7th Floor
P.O. Box 7921
Madison, WI 53707-7921
(608) 266-7718
http://www.dnr.state.wi.us/org/aw/air/

Wyoming

Wyoming Department of Environmental Quality
Air Quality Division
122 West 25th Street
Cheyenne, WY 82002
(307) 777-7391
http://deq.state.wy.us/aqd.htm

APCD = Air Pollution Control District; AQMD = Air Quality Management District.

Source: Elsevier Science Inc.

T. International Contacts

The Internet has numerous sources of information on international environmental laws and regulations. Two comprehensive resources are discussed below, but these are only the tip of the iceberg. For Web links to other international environmental agencies, refer to the STAPPA/ALAPCO Web links for international organizations (http//www.4cleanair.org/international.html).

INFOTERRA/USA

INFOTERRA is an international environmental referral and research network made up of about 175 countries. It is coordinated by the United Nations Environment Programme in Nairobi, Kenya. The U.S. focal point for INFOTERRA (INFOTERRA/USA), which is located at EPA headquarters in Washington, DC, is managed by EPA's Office of Information Resources Management. The office assists U.S. residents and the international community in obtaining international environmental information. It also publishes Earthlink, an international environmental newsletter.

INFOTERRA/USA is located at EPA, 3404, 401 M St., SW, Washington, DC 20460, (202) 260-5917. It can be reached online through e-mail: library-infoterra@epamail.epa.gov or at http://www.epa.gov/earlink1/INFOTERRA/index.html.

You can request that Earthlink be automatically e-mailed to you via EPA's public listservers service. Instructions on how to subscribe to this free service are available via http://www.epa.gov/epahome/listserv.htm.

Consortium for International Earth Science Information Network

The Consortium for International Earth Science Information Network maintains a Web site containing a plethora of U.S. and international environmental regulatory information. The site can be accessed at http://www.ciesin.org.

U. Associations

Many state and industry associations have compiled comprehensive regulatory and technical guidance resources. Some of the more well-known associations have developed Web sites that contain helpful information:

- *STAPPA/ALAPCO*—STAPPA and ALAPCO are two national groups representing air pollution control officials in the 55 states/territories and 165 major metropolitan areas in the United States. STAPPA/ALAPCO operates a Web site (http://www.4cleanair.org/about.html) that provides a wealth of information. The organization can be contacted via the Web site or by calling (202) 624-7864.
- *Air & Waste Management Association*—A&WMA can be contacted via http://www.awma.org or by calling (412) 232-3444.
- *American Institute of Chemical Engineers (AIChE)*—AIChE is one of the most well-known chemical engineering professional organizations. Further information is available via http://www.aiche.org or by calling (800) 242-4363.
- *Emissions Marketing Association (EMA)*—EMA promotes market-based trading solutions for environmental control by fostering member access to traders, brokers, and utilities seeking to buy, sell, and trade their emission allowances, credits, offsets, etc., for SO_2, NO_x, and other regulated emissions. The association can be contacted via http://www.emissions.org or by calling (414) 276-3819.

For further contacts, visit the Chemical Industry home page, which offers Web links to various chemical industry associations (http://www.neis.com/associations.html). In addition, the Air Pollution Control Resources Web site provides an extensive list of Web links to industry and trade association sources of air information (http://members.aol.com/CleenAir/apc/indus.htm).

V. Online Environmental News Information Sources

A number of useful environmental news information sources are now available online:

- *Capital Reports' News Digest* (http://www.caprep.com/new_dig3.htm), which provides frequently updated summaries of new EPA regulations and enforcement actions.
- EPA *Enviro-NewsBrief*, a daily update providing summaries of news articles pertaining to EPA and general governmental issues. Information on how to subscribe to this listserver can be found on the EPA Headquarters Information Resources Center Web site (http://www.epa.gov/natlibra/hqirc/enb.htm).
- *Internet NewsBrief*, a weekly update on new and/or useful environmental Internet resources (http://www.epa.gov/natlibra/hqirc/inb.htm).
- *EPA press releases*—Press releases issued by EPA can be accessed online via http://www.epa.gov/epahome/press.htm.

W. Newsletters and Journals

A number of helpful air-related publications are available. Table 9-13 lists select clean air publications in print, including the publisher, frequency, and price.

TABLE 9-13
Clean Air Publications

Title	*Publisher*	*Frequency*	*Price*
The Air Pollution Consultant	Elsevier Science Inc. 8773 South Ridgeline Blvd. Highlands Ranch, CO 80126 (303) 470-1900, ext. 12 gfsmith@elsevier-hr.com	Bimonthly (7 issues/year)	$775/yr
Air Pollution Monitoring Newsletter	McIlvaine Company 2970 Maria Ave. Northbrook, IL 60062 (847) 272-0010	Monthly	$270/yr
Air Pollution Regulatory Analysis Service	Elsevier Science Inc. 8773 South Ridgeline Blvd. Highlands Ranch, CO 80126 (303) 470-1900, ext. 12 gfsmith@elsevier-hr.com	Within ten business days after a proposed or final rule affecting stationary sources is published in the *FR*, a detailed analysis of the regulation is mailed to subscribers	$770/yr
Air/Water Pollution Report's Environment Week	Business Publishers, Inc. 951 Pershing Drive Silver Spring, MD 20910-4464 (800) 274-6737	Weekly	$867/yr
Atmospheric Environment	Elsevier Science Inc. Customer Support Department 655 Avenue of the Americas New York, NY 10010 (212) 633-3730 usinfo-f@elsevier.com	Biweekly	$3,561/yr
Clean Air Report	Inside Washington Publishers P.O. Box 7167 Ben Franklin Station Washington, DC 20044 (800) 424-9068 (703) 416-8500 in Washington DC	Biweekly	$645/yr
Journal of the Air & Waste Management Association (A&WMA)	A&WMA One Gateway Center, 3rd floor Pittsburgh, PA 15222 (412) 232-3444	Monthly	$240/yr

FR = Federal Register.

Source: Elsevier Science Inc.

10 Road Maps for Navigating Key EPA Web Sites

The Internet has experienced a meteoric rise in popularity and become a useful tool for accessing an incredible amount of information. To facilitate rapid information transfer, EPA has created a number of Web sites that provide CAA information, such as the Technology Transfer Network (TTN) Web sites discussed in Chapter 9 (page 9.1). As a result, you can now access EPA data, regulations, technical guidance documents, and countless other air-related resources directly from your personal computer.

Unfortunately, this virtually unlimited access to information can sometimes be overwhelming, especially if you don't have specific addresses for the Web sites you want to access. How many times have you spent hours looking for air-related information that you know *must* be available online, but you just can't find? To end this frustration, we have compiled road maps to several key EPA air-related Internet sites, including seven TTN Web sites, the Office of Air & Radiation (OAR) Web site, and the Office of Air Quality Planning and Standards (OAQPS) Web site. Table 10-1 lists the road maps and indicates where they can be found in this chapter. Individual site road maps are then presented in Tables 10-2 through 10-11. For each road map, we include a snapshot of the actual Web page followed by a breakdown of what each link offers (and its corresponding "address"). In most cases, Web page side bars have not been explained because their contents are usually self-explanatory based on their title (e.g., a link entitled "Contacts" will contain a list of EPA contacts). Use of these road maps should make your information searches smoother, especially when used in conjunction with the resources provided in Chapter 9 and the Internet directory found in Chapter 11.

As with any Internet information, please note that Web sites are constantly changing. As a result, the uniform resource locator (URL), i.e., the "address" of a Web site, may change from the listing provided in this guide. (If this occurs and the site still exists, it should be fairly easy, using the search engines listed in Table 11-1, to locate the new URL. APC

TABLE 10-1
List of Road Maps Provided in This Chapter

Web site	*Contents*	*Road map page number*
AIRS—Aerometric Information Retrieval System	Access to a computer database containing information about criteria air pollutants across the United States	10.3
AMTIC—Ambient Monitoring Technology Information Center	Information on ambient air monitoring programs, monitoring methods, air quality trends, nonattainment areas, and federal ambient air quality monitoring regulations	10.5
CHIEF—Clearinghouse for Inventories and Emission Factors	EPA's latest information and tools for estimating air pollutant emissions and performing air emission inventories	10.7
CATC—Clean Air Technology Center	Technical resources on emerging and existing air pollution prevention and control technologies, including: • Access to the RACT/BACT/LAER Clearinghouse, which contains data on prevention and control technology determinations made by states and local permitting agencies; this information can be used to help identify appropriate technologies for mitigating or treating most air pollutant emissions • Technical reports, cost information, and software	10.9
EMC—Emission Measurement Center	Guidance on emission testing and monitoring methods	10.11
OAQPS—Office of Air Quality Planning and Standards	Links to sites containing information regarding all OAQPS programs and regulations	10.13
OAR—Office of Air & Radiation	Information on OAR's responsibilities, contacts, links to various OAR Web sites, regulations, access to OAR publications, etc.	10.15
OAR P&G—OAR Policy & Guidance Documents	Access to recent and historic rules, policy, and guidance documents issued by OAR	10.17
SCRAM—Support Center for Regulatory Air Models	Information on atmospheric dispersion models that support CAA regulatory programs, including: • EPA modeling project data • Air quality dispersion models and user's guides • Meteorological data • Modeling guidance, tutorials, and other useful information • Lists of EPA and state modeling contacts	10.18
UATW—Unified Air Toxics Website	HAP information clearinghouse, which includes information on: • EPA rules concerning HAPs (e.g., MACT, NESHAP regulations); CAA Section 112 requirements, studies and reports; HAP policy, initiatives, and guidance • List of HAPs and information on HAP sources • Information on how to access technical HAP information via EPA hotlines, the Internet, and technical documents • Overview of the CAA Section 112 air toxics provisions • Lists of EPA, state, and local agency contacts	10.19

BACT = best available control technology; HAP = hazardous air pollutant; LAER = lowest achievable emission rate; MACT = maximum achievable control technology; NESHAP = national emission standards for hazardous air pollutants; RACT = reasonably available control technology.

Source: Elsevier Science Inc.

TABLE 10-2
Road Map of the AIRS Web Site

http://www.epa.gov/airs

Page link	*Description*	*Address*
Plant Information	Estimated annual emissions of criteria pollutants, VOCs, and total particulate matter from individual major sources. Reports can be generated for the entire United States, EPA regions, or individual states by specific pollutant. Depending on the report criteria selected, the reports will show 1) ranking of source emissions, 2) compliance status, 3) facility addresses, 4) number of sources emitting certain pollutants, 5) total pollutant emissions by SIC code, and 6) pollutant emissions reported by year.	http://www.epa.gov/airsweb/sources.htm
Monitor Statistics	• The Monitor Reports Web page (http://www.epa.gov/airsweb/monreps.htm) provides annual summaries of criteria pollutant data from individual monitoring stations across the United States. These summaries include information such as the highest values reported, the average value, and the number of values above NAAQS. • The Monitor Queries site (http://www.epa.gov/aqspub11/select.html) offers direct access to an interim AIRS database that includes air quality data for the last 15 years. It offers information on a wider range of pollutants and data elements than provided in the Monitor Reports.	http://www.epa.gov/airsweb/monitors.htm
Maps	Indicates the location of major air pollution sources, monitoring sites, and areas of the country where air pollution levels exceed EPA health-based standards (criteria pollutants only). Each map provides information about a single pollutant at the national scale.	http://www.epa.gov/airsweb/maps.htm
Contacts	Names, addresses, and phone numbers of EPA, state, and local environmental agency air pollution contacts.	http://www.epa.gov/airsweb/contacts.htm
Technical Users	Designed for use by state and local agencies when collecting, validating, and storing air pollution data in the AIRS database.	http://www.epa.gov/ttn/airs
AIRS Database Basic Facts	Helpful overview of the AIRS database and access information.	http://www.epa.gov/airs/airs2.html

TABLE 10-2
Road Map of the AIRS Web Site—Continued
http://www.epa.gov/airs

Page link	*Description*	*Address*
Publications Using AIRS Data	Air quality trends and emissions trends reports created using AIRS data.	http://www.epa.gov/airs/publicat.html
AIRS Data Websites	Link to the AIRSWeb, which provides easy public access to certain AIRS data including 1) criteria pollutant ambient air quality for the last 5 years from 4,000 air monitoring sites nationwide, and 2) criteria air pollutant, total particulate matter, and volatile organic compound emissions and regulatory compliance status for 9,000 EPA-regulated major point sources. Reports are generated on demand using data that are updated monthly.[1]	http://www.epa.gov/airsweb
AIRS Data Contacts	State and local contacts responsible for compiling air pollution data and storing them in the AIRS database.	http://www.epa.gov/airsweb/contsl.htm
AIRS Database Products	Summary of methods that can be used to access AIRS data: the AIRSWeb, AIRS personal computer software, and an upcoming AIRS Graphics on the Web (AGWeb) site that will allow access to additional AIRS reports, charts, and maps.	http://www.epa.gov/airs/products.html

AIRS = Aerometric Information Retrieval System; NAAQS = national ambient air quality standards; SIC = Standard Industrial Classification; VOC = volatile organic compound.

[1]At this time, the AIRSWeb site only provides a small part of the information contained in the voluminous AIRS database. EPA plans to make all of the AIRS data available on the Internet in the near future.

Source: Elsevier Science Inc.

TABLE 10-3
Road Map of the AMTIC Web Site
http://www.epa.gov/ttn/amtic

Page link	*Description*	*Address*
Monitoring Programs		
PM-2.5 Monitoring	PM-2.5 regulations, reference methods, contacts, policy and guidance, quality assurance information, training opportunities, etc.	http://www.epa.gov/ttn/amtic/amticpm.html
PAMS	Information on the photochemical assessment monitoring stations (PAMS) program, including contacts	http://www.epa.gov/ttn/amtic/pamsmain.html
Visibility (IMPROVE)	Details on the monitoring network established by IMPROVE to collect visibility data associated primarily with Class I areas	http://www.epa.gov/ttn/amtic/visdata.html
QA/QC	QA/QC control documents, National Performance Audit Program information, Precision and Accuracy Reporting System (PARS) details, QA/QC contacts, and QA Handbook information	http://www.epa.gov/ttn/amtic/qaqc.html
GEMS/AIR (WHO)	Information on the GEMS/AIR global urban air pollution monitoring and assessment program, which is operated by the World Health Organization (WHO) and the United Nations Environment Programme (UNEP)	http://www.epa.gov/ttn/amtic/gems.html
Saturation Monitoring	Saturation monitoring information	http://www.epa.gov/ttn/amtic/satmon.html
Air Superfund	Compendium of Air Superfund Documents, Air Superfund Fact Sheets, and other information	http://www.epa.gov/ttn/amtic/superfnd.html
Monitoring Methods		
Criteria Pollutants	Designated EPA reference and equivalent monitoring methods for criteria pollutants and related documents	http://www.epa.gov/ttn/amtic/criteria.html
Air Toxics	EPA monitoring methods and information on sampling/analysis of air toxics (e.g., HAPs)	http://www.epa.gov/ttn/amtic/airtox.html
Open Path	Proposed open-path siting criteria and other open-path analyzer information	http://www.epa.gov/ttn/amtic/longpath.html
Pollutant Standard Index	Pollutant Standard Index prototype program	http://www.epa.gov/ttn/amtic/psi.html

TABLE 10-3
Road Map of the AMTIC Web Site—Continued
http://www.epa.gov/ttn/amtic

Page link	*Description*	*Address*
Documents & Articles		
Technical Guidance	Technical guidance documents available for downloading	http://www.epa.gov/ttn/amtic/cpreldoc.html
Current Articles	Lists of recent articles of interest related to ambient monitoring	http://www.epa.gov/ttn/amtic/info/curartcl.txt
ORD/NERL Documents	Downloadable files containing documents issued by EPA's Office of Research and Development (ORD), National Exposure Research Laboratory (NERL), etc.	http://www.epa.gov/ttn/amtic/dira.html
Trends & Nonattainment	• Documents and graphics associated with air quality and emissions trends reports • Air quality and emissions trends and nonattainment links	http://www.epa.gov/ttn/amtic/trends.html
Regulations, Policy and Guidance	• Ambient air monitoring regulations contained in 40 *CFR* Parts 50, 53, and 58 • Proposed ambient monitoring regulations • Recently promulgated ambient monitoring regulations • Proposed PM and ozone NAAQS information • Policy and guidance information	http://www.epa.gov/ttn/amtic/codefed.html

AMTIC = Ambient Monitoring Technology Information Center; HAP = hazardous air pollutant; NAAQS = national ambient air quality standards; PM = particulate matter; PM-2.5 = particulate matter with an aerodynamic diameter of less than or equal to 2.5 microns; QA/QC = quality assurance/quality control.

Source: Elsevier Science Inc.

TABLE 10-4
Road Map of the CHIEF Web Site[1]
http://www.epa.gov/ttn/chief

Page link	*Description*	*Address*
Emission Inventories		
Emission Trends Report	Access to the most recent *National Air Pollutant Emissions Trends Report*	http://www.epa.gov/oar/emtrnd
Inventory Data & Projection	National emission inventory data and projections, including access to the current *National Air Quality Trends Report* and Ozone Transport Commission NO_x data	http://www.epa.gov/ttn/chief/ei_data.html
Emission Inventory Guidance	Emission inventory guidance, including EPA's general inventory guidance, information on criteria pollutants, and source classification codes	http://www.epa.gov/ttn/chief/ei_guide.html
1996 Emission Inventory	Central information source containing guidance to assist state and local agencies in preparing their Clean Air Act Amendments of 1990-mandated 1996 emission inventories (required to be submitted to EPA by July 15, 1998)	http://www.epa.gov/oar/oaqps/efig/ei
Emission Inventory Improvement Program	Information on the Emission Inventory Improvement Program's efforts to improve the quality of emissions information and develop systems for collecting, calculating, and reporting emissions data	http://www.epa.gov/oar/oaqps/eiip
Emission Factors		
AP-42 & Other References	• AP-42 Fifth Edition (by chapter), including AP-42 supplements, final background documents, and draft sections under review • L&E documents (discussed below) • Potentially useful emission estimation information • Fax CHIEF, a means of accessing AP-42, current and back issues of the *CHIEF Newsletter*, and several other technical publications via fax (see page 9.7 for more information on Fax CHIEF) • Source Classification Codes	http://www.epa.gov/ttn/chief/ap42etc.html

TABLE 10-4
Road Map of the CHIEF Web Site[1]—Continued

http://www.epa.gov/ttn/chief

Page link	*Description*	*Address*
Emission Estimation Software	• Air CHIEF CD-ROM • FIRE • Fugitive dust emissions estimation software (WIND and MECH) • Inventory projection software • Landfill Air Emissions Estimation Model • National Emission Trends CD-ROM • PCBEIS • PM CALC • SPECIATE • TANKS • WATER8 • CHEMDAT8 • Links to other software resources • A more detailed list of available emission estimation software is presented in Table 9-3 on page 9.8.	http://www.epa.gov/ttn/chief/software.html
L&E's (Air Toxics)	Access to final L&E documents that characterize toxic emissions from certain source categories, including general descriptions of the emitting processes, potential release points, and emission factors (see Table 9-4 on page 9.11 for a complete list of available L&E documents)	http://www.epa.gov/ttn/chief/ap42etc.html#LE

CHIEF = Clearinghouse for Inventories and Emission Factors; HAP = hazardous air pollutant; L&E = locating and estimating; NO_x = nitrogen oxides.

[1]The CHIEF Web site's sidebar (not shown above) also includes several useful links. The EFIG (Emission Factor and Inventory Group) Info link provides access to EPA emission inventory contacts, *CHIEF Newsletters,* the Fax CHIEF document index, current emission factor and inventory guidance, EPA memos and letters regarding emission inventories and factors, instructions on how to subscribe to the CHIEF List Server, etc. The Related Sites link ties into other valuable Web sites and online access to several key EPA documents (e.g., national emission and air quality trend reports, the Clean Air Act, the *EPA Green Book,* and the *Health Effects Notebook for HAPs*).

Source: Elsevier Science Inc.

TABLE 10-5
Road Map of the CATC Web Site
http://www.epa.gov/ttn/catc

Page link	*Description*	*Address*
About CATC	Overview of the contents of the CATC Web site.	http://www.epa.gov/ttn/catc/about.html
What's New	List of what's new on the CATC Web site.	http://www.epa.gov/ttn/catc/new.html
News & Bulletins	Access to issues of the CATC newsletter.	http://www.epa.gov/ttn/catc/news.html
Contact CATC	Resources for obtaining further information on CATC.	http://www.epa.gov/ttn/catc/contact.html
Search CATC	Search engine for the CATC and/or other TTN Web sites.	http://www.epa.gov/ttn/ttn_search.html
CATC Product Information	• Air pollution technical reports that outline industry-specific emissions and/or review control/management technologies. • Access to chapters of the *Air Pollution Control Costs Manual*, air pollution control costs escalation indices, and a control equipment cost spreadsheet. • Software and associated manuals, including the Landfill Gas Emissions Model, the Air Compliance Advisor Model, the HAP-PRO Model, the Solvent Alternative Guide Model, the Electrostatic Precipitator Model, and the Integrated Air Pollution System Program. • RBLC software and user manuals. • CATC newsletters.	http://www.epa.gov/ttn/catc/products.html
Technology & Regulation Databases	Access to the RBLC, which is designed to provide state and local agencies with information for making RACT, BACT, and LAER determinations. However, owners/operators can also use the RBLC to research candidate technologies likely to qualify for RACT, BACT, or LAER. The searchable Web site contains information distilled from air permits submitted by most state and local agencies in the United States on facilities for which RACT, BACT, and LAER determinations have been made. It contains more than 3,500 determinations with information on over 200 pollutants and 1,000 industrial processes. RBLC also contains a database that summarizes all emission standards and control technique guidelines issued by EPA.	http://mapsweb.rtpnc.epa.gov/RBLCWeb/b102.htm

TABLE 10-5
Road Map of the CATC Web Site—Continued

http://www.epa.gov/ttn/catc

Page link	Description	Address
CICA US-Mexican Border	Link to the CICA (Centro de Infomación Sobre Contaminación de Aire) Web site, which provides technical information regarding air pollution along the U.S./Mexico border.	http://www.epa.gov/ttn/catc/cica
Small Business Assistance	Link to the Small Business Assistance Program Web site, which provides technical assistance for small businesses that need to comply with the CAA.	http://www.epa.gov/ttn/sbap

BACT = best available control technology; CATC = Clean Air Technology Center; LAER = lowest achievable emission rate; RACT = reasonably available control technology; RBLC = RACT/BACT/LAER Clearinghouse; SBAP = Small Business Assistance Program; TTN = Technology Transfer Network.

Source: Elsevier Science Inc.

TABLE 10-6
Road Map of the EMC Web Site
http://www.epa.gov/ttn/emc

Page link	*Description*	*Address*
FAQ	Frequently asked questions (FAQS) about emission test methods, laboratories, performance specifications, and other emission measurement issues	http://www.epa.gov/ttn/emc/facts.html
Methods	Guidance on the use of the following emission test methods (including lists of methods): • Methods proposed or promulgated in the *Federal Register* • Source category-approved alternative methods (i.e., EPA-approved alternatives to test methods required by 40 *CFR* Parts 60, 61, and 63) • Conditional methods (methods that have been evaluated by EPA and may be applicable to certain source categories) • Preliminary methods (methods that may be useful in limited applications even though further information is needed) • "Idea box" methods (methods that may be reviewed by EPA in the future)	http://www.epa.gov/ttn/emc/tmethods.html
Monitoring	Technical guidance on performance specifications, continuous emission monitoring systems, compliance assurance monitoring, FTIR technologies, and acid rain monitoring	http://www.epa.gov/ttn/emc/monitor.html
What's New?	Updates on new documents available on the EMC Web site	http://www.epa.gov/ttn/emc/new.html
Who is EMC?	Description of the EMC, including a staff directory	http://www.epa.gov/ttn/emc/who.html
Technical Support	Information on how to obtain technical emissions testing information from EPA	http://www.epa.gov/ttn/emc/email.html
Related Web Sites	Links to other Web sites that contain emission measurement information	http://www.epa.gov/ttn/emc/related.html

TABLE 10-6
Road Map of the EMC Web Site—Continued
http://www.epa.gov/ttn/emc

Page link	*Description*	*Address*
Instructional Material	• EMC software (databases and other supporting software) • List of available EMC videos • List of upcoming EMC meetings, conferences, and workshops • Guideline documents that clarify undefined technical areas of methods and performance specifications • Information documents that provide technical information used by regulatory agencies to make decisions about acceptance of data and alternative methods/monitoring • News items pertaining to emission measurement or emission standards • Office of Research and Development documents	http://www.epa.gov/ttn/emc/in_mat.html
Other EPA Web Sites	Links to other related EPA Web sites	http://www.epa.gov/ttn/emc/other.html

EMC = Emission Measurement Center; FTIR = fourier transform infrared spectroscopy.

Source: Elsevier Science Inc.

TABLE 10-7
Road Map of the OAQPS Web Site
http://www.epa.gov/oar/oaqps

Page link	*Description*	*Address*
Air Quality	Links to sites containing information on OAQPS efforts to: • Set air quality standards • Monitor air quality • Collect air quality data • Report air quality trends • List nonattainment areas • Inform the public about local air quality • Help states develop SIPs • Participate in cooperative efforts with other agencies and organizations	http://www.epa.gov/oar/oaqps/cleanair.html
Visibility	• Overview of the visibility impairment problem • Long-term visibility trends • Long-term visibility monitoring program description • Summary of EPA's regional haze program • Descriptions of other visibility programs, including the National Park Service Visibility Program, the Grand Canyon Visibility Transport Commission, and the Western Regional Air Partnership • Links to CAA sections pertaining to visibility	http://www.epa.gov/oar/vis/overview.html
Technical Info	Link to the TTN Web site	http://www.epa.gov/ttn
Air Toxics	Link to the Unified Air Toxics Website (UATW); a road map of the UATW begins on page 10.19	http:/www.epa.gov/ttn/uatw
Overview	Brief overview of OAQPS responsibilities and main projects; for more details see Projects below	http://www.epa.gov/oar/oaqps/overview.html
Education	Links to EPA and other air pollution training Web sites	http://www.epa.gov/oar/oaqps/educatn.html
Emissions	Links to sites containing information on: • Air pollutant emission trends • A schedule of upcoming OAQPS regulations • Emission factors and inventories • Emissions testing	http://www.epa.gov/oar/oaqps/emissns.html
New Actions	Access to CAA policy, guidance, and regulations issued in the previous six months	http://www.epa.gov/ttn/oarpg/ramain.html
Website Info	Information on who to contact for information, questions, comments regarding the OAQPS Web site	http://www.epa.gov/oar/oaqps/website.html
Organization	• OAQPS organizational chart • Links to OAQPS division Web sites • List of OAQPS senior management contacts	http://www.epa.gov/oar/oaqps/org/index.html#orgchart

TABLE 10-7
Road Map of the OAQPS Web Site—Continued
http://www.epa.gov/oar/oaqps

Page link	*Description*	*Address*
Projects	Useful links to a number of Web sites containing information on the following OAQPS project areas: • Air pollution monitoring • Air pollution emissions • Air pollution data • Air quality monitoring • Inter-agency programs • Conferences, workshops, and training	http://www.epa.gov/oar/oaqps/projects.html
Regulations	Links to information on: • Estimated dates for completion of OAQPS regulations • Air toxics rules • The CAA • The operating permit program • Accessing the *Federal Register* online • The final pulp and paper rule • Region V SIP documents searchable database	http://www.epa.gov/oar/oaqps/regulatn.html
Publications	Access to a variety of useful online OAQPS publications, including AP-42, air quality trends reports, emission trends reports, newsletters, etc.	http://www.epa.gov/oar/oaqps/publicat.html
Assistance	List of many helpful information resource links to software/data, technical assistance Web sites, organizations/associations Web sites, training/education resources, EPA fact sheets, EPA regional Web sites, and non-EPA environmental Web information sources	http://www.epa.gov/oar/oaqps/assist.html
New	Updates on information recently added to the OAQPS Web site	http://www.epa.gov/oar/oaqps/oaqnew.html
Upcoming	Links for the current month to recent and upcoming OAQPS regulatory actions (e.g., proposed rules, final rules), including contact information, and a list of recent and upcoming major *Federal Register* notices	http://www.epa.gov/oar/oaqps/upcoming.html
Search	Search engine that allows the user to search all of the OAR Web site, the TTN Web site, and/or the OAR Policy & Guidance Documents Web site	http://www.epa.gov/oar/oarsrch.html
Index	Alphabetical index of keywords in the title of OAR Web pages; associated links are provided	http://www.epa.gov/oar/keywords/index.html
Public Forum	Open forum that allows anyone to submit questions, comments, and responses about air pollution	http://www.epa.gov/cgi-bin/netforum/oaqps/a/1
Comments	Form for submitting comments to OAR regarding its Web pages	http://www.epa.gov/cgi-bin/oarcmnt.pl

OAQPS = Office of Air Quality Planning and Standards; OAR = Office of Air & Radiation; SIP = state implementation plan; TTN = Technology Transfer Network.

Source: Elsevier Science Inc.

TABLE 10-8
Road Map of EPA's OAR Web Site

http://www.epa.gov/oar

Page link	*Description*	*Address*
Ozone Layer	Link to EPA's Ozone Depletion Web page, which contains information about the science of ozone depletion, EPA regulations designed to protect the ozone layer, ozone-depleting substance substitutes, methyl bromide, etc.	http://www.epa.gov/ozone
Urban Air	Link to EPA's Office of Air Quality Planning and Standards (OAQPS) home page; for a road map to the OAQPS site, turn to page 10.13	http://www.epa.gov/oar/oaqps
Radiation	The OAR Radiation Protection Division's home page designed to provide information on radiation risks, clean up of radioactively contaminated sites, mixed waste, air standards, radon, etc.	http://www.epa.gov/radiation
Acid Rain	EPA's Acid Rain home page, which provides details on the acid rain program, environmental effects of acid rain, SO_2 emission trading, the NO_x reduction program, emission monitoring and reporting, and conservation and renewable energy incentives	http://www.epa.gov/acidrain/ardhome.html
Vehicles/ Engines	The OAR Office of Mobile Sources home page, which contains information on emissions from motor vehicles, fuels, and nonroad engines	http://www.epa.gov/OMSWWW/omshome.htm
Air Toxics	The Unified Air Toxics Website (UATW), a comprehensive source of information on regulations, EPA guidance, contacts, etc. regarding air toxics; a road map of the UATW is shown on page 10.19.	http://www.epa.gov/ttn/uatw
Partners	Links to other federal, state, and local agencies with air pollution programs	http://www.epa.gov/oar/partners.html
Prevention	EPA's Atmospheric Pollution Prevention Division's Web page, which provides access to information on improving energy efficiency, developing better chlorofluorocarbon (CFC) substitutes, and capturing and using methane releases instead of emitting them to the atmosphere	http://www.epa.gov/appd.html
Indoor Air	The Indoor Air Quality home page, containing resources on radon, formaldehyde, asbestos, lead, and other indoor air pollutants	http://www.epa.gov/iaq

TABLE 10-8
Road Map of EPA's OAR Web Site—Continued

http://www.epa.gov/oar

Page link	Description	Address
Visibility	Information on visibility problems, trends, and programs designed to improve visibility	http://www.epa.gov/oar/vis/overview.html
Basic Facts	• General overview of OAR's responsibilities • Access to a .pdf version of the *OAR Information Resources* document, which lists Internet resources, hotlines, written publications, and other resources pertaining to air pollution topics	http://www.epa.gov/oar/oarfacts.html
Organization	List of OAR contacts and their telephone numbers	http://www.epa.gov/oar/oarofcs.html
Resources	Links to other useful Internet resources	http://www.epa.gov/oar/oarrsrc.html
Mission	Description of OAR's mission, vision, and values	http://www.epa.gov/oar/oarmisn.html
AIRLinks	What's new this week at OAR, breaking news, and hot OAR topic information	http://www.epa.gov/airlinks
Regulations	• Access to air laws and OAR regulations, including CAA documents, operating permit regulations, air toxics rules, ozone depletion regulations, and the final pulp and paper cluster rule • Link to the What's Upcoming in OAR Web site, which lists upcoming regulatory actions, major reports, important *Federal Register* notices, and a schedule of meetings • Link to the *Federal Register* Online Web site	http://www.epa.gov/oar/oarregul.html
Publications	Links to publications available on the OAR Web site	http://www.epa.gov/oar/oarpubs.html
Special Projects	• Reports and publications concerning OAR's hot topics (e.g., the *Air Quality Trends Report*), the OAR Implementation Plan, etc. • Link to the What's Upcoming in OAR Web site	http://www.epa.gov/oar/oario.html

OAR = Office of Air and Radiation; NO_x = nitrogen oxides; SO_2 = sulfur dioxide.

Source: Elsevier Science Inc.

TABLE 10-9

Road Map of the OAR P&G Web Site

http://www.epa.gov/ttn/oarpg

Page link	*Description*	*Address*
What's New	List of new material posted on the OAR P&G Web site	http://www.epa.gov/ttn/oarpg/new.html
Recent Actions	Quick and easy access to OAR policy, guidance, and regulations issued in the previous six months[1]	http://www.epa.gov/ttn/oarpg/ramain.html
Actions Sorted by CAA Title	Access to proposed and final rules, policy, and guidance documents categorized by CAA Title[2]:	www.epa.gov/ttn/oarpg/amend.html
	• Attainment/Maintenance of NAAQS (Title I)	http://www.epa.gov/ttn/oarpg/t1main.html
	• Mobile Sources (Title II)	http://www.epa.gov/ttn/oarpg/t2main.html
	• Hazardous Air Pollutants (Title III)	http://www.epa.gov/ttn/oarpg/t3main.html
	• Acid Deposition (Title IV)	http://www.epa.gov/ttn/oarpg/t4main.html
	• Operating Permits & New Source Review (Title V)	http://www.epa.gov/ttn/oarpg/t5main.html
	• Stratospheric Ozone (Title VI)	http://www.epa.gov/ttn/oarpg/t6main.html
	• Enforcement (Title VII)	http://www.epa.gov/ttn/oarpg/t7main.html
News and Bulletins	Miscellaneous news/bulletins available for downloading, including Carol Browner's House of Representatives testimony on several issues	http://www.epa.gov/ttn/oarpg/news.html
General Documents	Various documents available for downloading including: *Directory of OAQPS Information Services*, U.S./Canada Air Quality Agreement, Full Text of the CAA, Full Text of CAA Amendments, General Overview of the CAA	http://www.epa.gov/ttn/oarpg/gener.html
Clean Air Updates	List of upcoming OAR regulatory actions and major *Federal Register* publications (updated monthly)	http://www.epa.gov/ttn/oarpg/update.html
Search OAR P&G	TTN Web search engine that allows the user to select the TTN Web area(s) (e.g., OAR P&G) to be searched	http://www.epa.gov/ttn/ttn_search.html
Site Index	Lists and provides links to all Web pages that are part of the OAR P&G site; the index can be viewed as a hierarchical tree structure of the entire site or a list of all pages on the site in alphabetical order	http://www.epa.gov/ttn/oarpg/siteindx.html

NAAQS = national ambient air quality standards; OAQPS = Office of Air Quality Planning and Standards; OAR = Office of Air & Radiation; P&G = Policy and Guidance; TTN = Technology Transfer Network.

[1]This information also appears under the appropriate Title of the CAA on the "Actions Sorted by CAA Title" Web page.

[2]Information includes: 1) *Federal Register* notices (proposed and final preambles/rules, revision/clarification notices, supplemental notices, extension notices), 2) reports (reports to Congress, regulatory impact analyses, background information documents, enabling documents, staff papers, criteria documents), 3) memoranda (policy and guidance memos, white papers), and 4) fact sheets.

Source: Elsevier Science Inc.

TABLE 10-10
Road Map of the SCRAM Web Site
http://www.epa.gov/ttn/scram/main.htm

Page link	*Description*	*Address*
Homepage	Information on atmospheric dispersion models that support CAA regulatory programs	http://www.epa.gov/ttn/scram/main.htm
EPA Regional Modeling Center	Descriptions and data summaries associated with EPA modeling projects	http://www.epa.gov/ttn/scram/regmodcenter/rmcmain.htm
Free Software Utilities	Adobe Acrobat PDF Reader and zip/unzip programs available for downloading	http://www.epa.gov/scram001/software.htm
SCRAM Search	TTN search engine	http://www.epa.gov/ttn/ttn_search.html
SCRAM FTP	FTP download area for accessing certain modeling information	FTP://www.epa.gov/pub/scram001/scram
What's New	Announcements concerning additions and changes to models/programs, user's guides, guidelines, and support documents	http://www.epa.gov/scram001/t21.htm
Models	Files containing computer codes for various air quality dispersion models	http://www.epa.gov/scram001/t22.htm
User's Guides	User's guides for most of the models available from the SCRAM Web site	http://www.epa.gov/scram001/t23.htm
Model Change Bulletins	Comprehensive information on all changes made to the computer code for the air quality models and related programs offered on the SCRAM Web site	http://www.epa.gov/scram001/t24.htm
Met Data	Meteorological data including: 1) one or more years of surface data for various National Weather Service stations across the United States, and 2) mixing height data to be used in concert with surface data	http://www.epa.gov/scram001/t25.htm
Model Support	Modeling guidance, model clearinghouse, model tutorials, research abstracts, and other modeling information	http://www.epa.gov/scram001/t26.htm
Public Forum	State environmental links, state modeling contacts, other environmental links, frequently asked modeling questions, and EPA regional modeling contacts	http://www.epa.gov/scram001/t27.htm#10

FTP = file transfer protocol; PDF = portable document format; SCRAM = Support Center for Regulatory Air Models; TTN = Technology Transfer Network.

Source: Elsevier Science Inc.

TABLE 10-11
Road Map of the UATW[1]

http://www.epa.gov/ttn/uatw

Page link	*Description*	*Address*
Basic Facts	• Overview of information contained on UATW • Basic facts on HAPs and their health effects	http://www.epa.gov/ttn/uatw/basicfac.html
EPA Rules & Implementation	• Lists of final and future NESHAP MACT standards, including final or expected proposal/promulgating dates and EPA contact information • Generic MACT rule drafts • Coatings and composites coordinated rule development • MACT implementation strategy • Lists of HAPs and source categories • Miscellaneous HAP rules, policies, initiatives, and guidance on issues such as potential to emit, compliance assurance monitoring, etc. • List of control techniques guidelines • Access to Part 63 NESHAP regulations • Applicability Determination Index, which provides access to EPA regulatory interpretations	http://www.epa.gov/ttn/uatw/eparules.html
Pollutants & Sources	• List of HAPs • Definitions of HAP "major source" and "area source"	http://www.epa.gov/ttn/uatw/pollsour.html
Technical Resources	• List of HAP-related telephone hotlines • Summary of Internet technical information resources • List of technical documents concerning HAPs (including online access to some documents and ordering information for others)	http://www.epa.gov/ttn/uatw/techres.html
EPA Programs & Contacts	• Overview of the CAA air toxics provisions (Section 112) • EPA contacts, including the program contacts for each Section 112 subsection	http://www.epa.gov/ttn/uatw/epaprogs.html
State, Local and Tribal Programs	State, local, and tribal air quality program links	http://www.epa.gov/ttn/uatw/stprogs.html
STAPPA/ALAPCO	Link to the State and Territorial Air Pollution Program Administrators (STAPPA)/Association of Local Air Pollution Control Officials (ALAPCO) Web site, which is a very useful resource for air pollution links on a number of topics	http://www.4cleanair.org

TABLE 10-11
Road Map of the UATW[1]

http://www.epa.gov/ttn/uatw

Page link	Description	Address
What's New	Updates of new information available on the UATW	http://www.epa.gov/ttn/uatw/uatwnew.html
Search	Search engine to the UATW and/or other TTN Web sites	http://www.epa.gov/ttn_search.html.
Public Forum	Open forum for questions, comments, and responses concerning air toxics	http://www.epa.gov/cgi-bin/netforum/uatw/a/1
Site Map	Map of the UATW	http://www.epa.gov/ttn/uatw/sitemap.html

HAP = hazardous air pollutant; MACT = maximum achievable control technology; NESHAP = national emission standards for hazardous air pollutants; TTN = Technology Transfer Network; UATW = Unified Air Toxics Website.

[1]Air toxics are also referred to as hazardous air pollutants or toxic air pollutants.

Source: Elsevier Science Inc.

11 Directory of CAA-Related Web Sites

In addition to the Internet resources described in Chapters 9 and 10, there are a multitude of other useful CAA-related resources on the World Wide Web. To allow easy access to these Web sites, we have compiled a directory of air-related resources organized by keyword and site name. The directory, which is presented in Table 11-1, is the result of a comprehensive Internet search and is current as of October 1, 1998; however, it is by no means complete.

Due to the rapid pace of Internet expansion, Web sites are constantly changing (e.g., new sites are being added, old sites become obsolete). As a result, the URL (i.e., the "address" of a Web site) may change from the listing provided in this guide. (If this occurs and the site still exists, it should be fairly easy, using the search engines listed in the directory, to locate the new URL.

Changes in Internet addresses and additional sites for listing in future directories are welcome. Please send them to Joseph J. Bernosky, P.E., Associate Editor, *The Air Pollution Consultant*, Elsevier Science Inc., 8773 South Ridgeline Blvd., Highlands Ranch, CO 80126-2329, (303) 470-1900, extension 11, Fax (303) 470-5119, j.bernosky@elsevier.com. APC

TABLE 11-1
CAA Internet Directory[1]

Site name/keyword	*Address (URL)*
Accidental release prevention	http://members.aol.com/cleanair4/docs/rmp-112.htm#GUID
	http://response.restoration.noaa.gov/cameo/dr_aloha/RMPtools
	http://response.restoration.noaa.gov/chemaids/rmp
Acid rain	
Allowance tracking system	http://www.epa.gov/acidrain/atsdata2.html
Emission monitoring and reporting	http://www.epa.gov/acidrain/so2emis.html
EPA's acid rain Web site	http://www.epa.gov/docs/acidrain
NO_x reduction	http://www.epa.gov/acidrain/noxpa.html
Regulations, policy, and guidance documents	http://www.epa.gov/ttn/oarpg/t4main.html
SO_2 trading	http://www.epa.gov/acidrain/trading.html
Web links	http://epa.gov/acidrain/links.htm
	http://www.4cleanair.org/acidrain.html
Adobe Acrobat Reader	http://www.adobe.com/prodindex/acrobat/readstep.html
Aerometric Information Retrieval System	http://www.epa.gov/airs
Air & Waste Management Association	http://www.awma.org
Air Pollution Control Resources home page	http://members.aol.com/CleenAir/index.htm
Air Pollution Distance Learning Network	http://www.epa.gov/oar/oaqps/eog/apdln.html
Air quality dispersion models	http://www.epa.gov/scram
AIRNOW	http://www.epa.gov/airnow
ALTA VISTA search engine	http://www.altavista.digital.com
Ambient Monitoring Technology Information Center	http://www.epa.gov/ttn/amtic
American Institute of Chemical Engineers	http://www.aiche.org
American National Standards Institute	http://www.ansi.org
AP-42 (EPA emission factors)	http://www.epa.gov/ttn/chief/ap42.html
Applicability Determination Index	http://134.67.104.12/cfdocs/adiwww/adiwww.html-ssi
Asbestos, National Asbestos Registry System	http://www.epa.gov/earth100/records/i20076.html
Atmospheric pollution prevention program	http://www.epa.gov/appd.html
Audit policy interpretive guidance	http://es.epa.gov/oeca/apolguid.html
Border Information Center on Air Pollution	http://www.epa.gov/ttn/catc/cica
California Air Resources Board	http://www.arb.ca.gov/training
Training	http://www.arb.ca.gov/training
Center for Environmental Information and Statistics	http://www.epa.gov/ceisweb1/ceishome
ChemFinder	http://chemfinder.camsoft.com
Chemical and process engineering resources	http://www.neis.com/cpe_resources.html
Chemical data	
ChemFinder	http://chemfinder.camsoft.com
Chemical fact sheets	http://www.epa.gov/chemfact
Hazardous chemical database	http://ull.chemistry.uakron.edu/erd/sources.html
Chemical Industry Compliance Center	http://www.chemalliance.org

TABLE 11-1
CAA Internet Directory[1]—Continued

Site name/keyword	*Address (URL)*
Chemical Industry home page	http://www.neis.com
Chromium electroplating	http://www.epa.gov/ttn/uatw/csigc.html
Clean Air Act Policy Compendium	http://es.epa.gov/oeca/ore/aed/comp/index.html
Clean Air Act/Clean Air Act Amendments of 1990	
Full text	http://www.epa.gov/oar/oaq_caa.html
Plain English guide	http://www.epa.gov/oar/oaqps/peg_caa/pegcaain.html#index
Proposed legislation	http://members.aol.com/CleenAir/news/newslaws.htm
Proposed legislation	http://thomas.loc.gov/home/thomas2.html
Clean Air Marketplace Database	http://www.epa.gov/oar/opar/clam.html
Clearinghouse for Inventories and Emission Factors	http://www.epa.gov/ttn/chief
Code of Federal Regulations	
All parts	http://www.access.gpo.gov/nara/cfr/index.html
Title 40 only	http://www.epa.gov/epacfr40
Compliance assistance	
Chemical industry	http://www.chemalliance.org
Local governments	http://www.lgean.org
Metal finishing	http://www.nmfrc.org/nmfrchom.cfm
Paints and coatings	http://www.paintcenter.org
Transportation industry	http://www.transource.org
Pharmaceutical industry	http://206.4.171.30/rules/pharmact/index.cfm
Compliance information, stationary sources	http://es.epa.gov/oeca/ore/aed/comp/bcomp
Congressional Record	http://thomas.loc.gov/home/thomas2.html
Consortium for International Earth Science Information Network	http://www.ciesin.org
Control technologies	
Clean Air Marketplace Database (CLAM)	http://www.epa.gov/oar/opar/clam.html
Clean Air Technology Center (CATC)	http://www.epa.gov/ttn/catc
Cornell Environmental Law Institute	http://www.law.cornell.edu/topics/environmental.html
Courts	
Federal and state appellate court Web links	http://supct.law.cornell.edu/supct/directory/overview.html
Federal Courts Finder	http://www.law.emory.edu/FEDCTS
U.S. Court of Appeals for the District of Columbia Circuit	http://www.11.georgetown.edu/Fed-Ct/cadc.html
U.S. Court of Appeals for the First Circuit	http://www.law.emory.edu/1circuit
U.S. Court of Appeals for the Second Circuit	http://www.TouroLaw.edu/2ndCircuit
U.S. Court of Appeals for the Third Circuit	http://www.law.vill.edu/Fed-Ct/ca03.html
U.S. Court of Appeals for the Fourth Circuit	http://www.law.emory.edu/4circuit
U.S. Court of Appeals for the Fifth Circuit	http://www.ca5.uscourts.gov
U.S. Court of Appeals for the Sixth Circuit	http://www.law.emory.edu/6circuit
U.S. Court of Appeals for the Seventh Circuit	http://www.kentlaw.edu/7circuit
U.S. Court of Appeals for the Eighth Circuit	http://www.wulaw.wustl.edu/8th.cir

TABLE 11-1
CAA Internet Directory[1]—Continued

Site name/keyword	*Address (URL)*
Courts—Continued	
U.S. Court of Appeals for the Ninth Circuit	http://www.findlaw.com/casecode/courts/9th.html
U.S. Court of Appeals for the Tenth Circuit	http://www.law.emory.edu/10circuit
U.S. Court of Appeals for the Eleventh Circuit	http://www.law.emory.edu/11circuit
U.S. Supreme Court	http://supct.law.cornell.edu/supct/directory/united_states.html
Dry cleaners	
Inspection guidance and checklists	http://www.epa.gov/ttn/uatw/dcsigc.html
Reference manual	http://www.epa.gov/ttn/uatw/dryclean.pdf
Economic incentives	http://134.67.55.16:7777/aa/programs.nsf
Education and Outreach Group	http://www.epa.gov/oar/oaqps/eog
Electroplating	http://www.epa.gov/ttn/uatw/csigc.html
Elsevier Environmental Science & Technology	http://www.elsevier.enviroinfo.com/
Emission estimation software	http://www.epa.gov/ttn/chief/software.html
Emission factors	
AP-42	http://www.epa.gov/ttn/chief/ap42.html
Clearinghouse for Inventories and Emission Factors	http://www.epa.gov/ttn/chief
Emission inventories	
Clearinghouse for Inventories and Emission Factors	http://www.epa.gov/ttn/chief
EPA contacts	http://www.epa.gov/ttn/chief/contact2.html
	http://www.epa.gov/ttn/chief/reg-ei.txt
Guidance	http://www.epa.gov/ttn/chief/ei_guide.html
Emission Measurement Center	http://www.epa.gov/ttn/emc
Emissions Marketing Association	http://www.emissions.org
Emory Law Library	http://www.law.emory.edu/LAW/library.html
Enforcement	
Compliance information, stationary sources	http://es.epa.gov/oeca/ore/aed/comp/bcomp
OECA	http://www.epa.gov/oeca
OECA online newsletter	http://es.epa.gov/oeca/enforcement
Policy and guidance documents	http://www.epa.gov/ttn/oarpg/amend.html
Recent regulatory developments	http://members.aol.com/CleenAir/news/newsoeca.htm
Stationary source policies and guidance	http://es.epa.gov/oeca/ore/aed/comp/bcomp
Web links	http://www.4cleanair.org/enforce.html
ENVIROENE	http://es.epa.gov
Envirofacts	http://www.epa.gov/enviro/html/ef_home.html
Environmental Appeals Board	http://www.epa.gov/eab
Environmental law libraries	
Cornell Environmental Law Institute	http://www.law.cornell.edu/topics/environmental.html
Emory Law Library	http://www.law.emory.edu/LAW/library.html
Environmental Law World Wide Web site	http://www.webcom.com/~staber/rectinfo.html

TABLE 11-1
CAA Internet Directory[1]—Continued

Site name/keyword	*Address (URL)*
Environmental law libraries—Continued	
Federal court Web links	http://www.fjc.gov/WWWlinks/govlinks.html
LEXIS-NEXIS[2]	http://www.lexis-nexis.com/lncc/sources/libcont/envirn.html
Pace Virtual Environmental Law Library	http://joshua.law.pace.edu/env/vell6.html
U.S. House of Representatives Internet Law Library	http://law.house.gov
Washburn University School of Law	http://lawlib.wuacc.edu/washlaw/searchlaw.html
Environmental Law World Wide Web site	http://www.webcom.com/~staber/rectinfo.html
Environmental news online	
Capital Reports' News Digest	http://www.caprep.com/new_dig3.htm
EPA Enviro-NewsBrief	http://www.epa.gov/natlibra/hqirc/enb.htm
EPA press releases	http://www.epa.gov/epahome/press.htm
Internet NewsBrief	http://www.epa.gov/natlibra/hqirc/inb.htm
OECA enforcement newsletter	http://es.epa.gov/oeca/enforcement
Environmental Professional's Home Page	http://www.clay.net/ep.html
EPA document sources	
EPA Online Library System	http://www.epa.gov/natlibra/ols.htm
Government Printing Office access database	http://www.access.gpo.gov/su_docs/aces/aaces002.html
Guidance memos, Office of Air and Radiation	http://www.epa.gov/ttn/oarpg
National Center for Environmental Publications and Information	http://www.epa.gov/ncepihom/index.html
National Environmental Publications Information	http://www.epa.gov/cincl
National Technical Information Service	http://www.fedworld.gov/ntis/ntishome.html
EPA test method index	http://www.epa.gov/epahome/index
EPA	
Acid Rain Division	http://www.epa.gov/docs/acidrain
Atmospheric pollution prevention program	http://www.epa.gov/appd.html
Contacts, emission inventories	http://www.epa.gov/ttn/chief/reg-ei.txt
Contacts, Office of Enforcement	ftp://134.67.104.11/f-drive/sscd/txt/sscdcont.txt
EPA laboratories, Web links	http://www.nttc.edu/env/epalabs.html
EPA People Locator (all of EPA)	http://www.epa.gov:6706/natlocdcd/owa/locator.home
EPA People Locator (OAR only)	http://www.epa.gov/oar/oarpeopl.html
EPA Web page search engine	http://www.epa.gov/epahome/search.html
Government Information Locator Service	http://www.epa.gov/earth100/index.html
Home page	http://www.epa.gov
Hotlines	http://www.epa.gov/epahome/hotline.htm
Hotlines	http://www.epa.gov/ttn/uatw/techres.html#PHONE
Memoranda on applicability and compliance issues	http://134.67.104.12/cfdocs/adiwww/adiwww.html-ssi
Office of Air and Radiation	http://www.epa.gov/oar
Office of Air Quality Planning and Standards	http://www.epa.gov/oar/oaqps

TABLE 11-1
CAA Internet Directory[1]—Continued

Site name/keyword	Address (URL)
EPA—Continued	
Office of Atmospheric Programs	http://www.epa.gov/oar/oap.html
Office of Enforcement and Compliance Assurance	http://www.epa.gov/oeca
Office of Radiation and Indoor Air	http://www.epa.gov/radiation
Offices, labs, and regions	http://www.epa.gov/epahome/Offices.html
Offices, Web links	http://www.4cleanair.org/federal.html
Public list servers	http://www.epa.gov/epahome/listserv.htm
Regions, Web links	http://www.epa.gov/epahome/Offices.html
Regions, Web links	http://www.epa.gov/ttn/chief/relsites.html#epa
Semi-annual regulatory agenda	http://www.epa.gov/fedrgstr/unified.htm
TTN Web site	http://www.epa.gov/ttn
Excite search engine	http://www.excite.com
Factor Information and Retrieval Data System (FIRE)	http://www.epa.gov/ttn/chief/fire.html
Federal Courts Finder	http://www.law.emory.edu/FEDCTS
Federal Register	
EPA (October 1994 to present)	http://www.epa.gov/fedrgstr
Government Printing Office (January 1994 to present)	http://www.access.gpo.gov/su_docs/aces/aaces002.html
Government Printing Office, searchable index	http://www.access.gpo.gov/su_docs/aces/aces140.html
General environmental resources	
Air Pollution Control Resources home page	http://members.aol.com/CleenAir/index.htm
Chemical and process engineering resources	http://www.neis.com/cpe_resources.html
Chemical Industry home page	http://www.neis.com
Environmental Professional's home page	http://www.clay.net/ep.html
Environmental resources	http://www.neis.com/environmental_resources.html
Government databases and search engines	http://members.aol.com/CleenAir/apc/fed3.htm#AIRDBASE
Internet resources for the environmental scientist	http://www.imt.net/~dcouncil/env.html
WWW Virtual Library: Environment	http://earthsystems.org/Environment.html
Geographic/ecosystems initiatives	http://www.epa.gov/ttn/gei
Government Information Locator Service	http://www.epa.gov/earth100/index.html
Government Printing Office	
Database	http://www.access.gpo.gov/su_docs/aces/aaces002.html
Document orders	http://www.access.gpo.gov/su_docs/index.html
Ground level ozone	http://www.epa.gov/docs/region01/eco/ozone/hold/index.html
Hazardous air pollutants	
EPA technical documents	http://www.epa.gov/ttn/uatw/techres.html#TECHDOC
Health effects	http://www.epa.gov/ttn/uatw/hapindex.html
List of hazardous air pollutants	http://www.epa.gov/ttn/uatw/188polls.txt
List of source categories and MACT schedule	http://www.epa.gov/r02earth/air/airtox.htm

TABLE 11-1
CAA Internet Directory[1]—Continued

Site name/keyword	*Address (URL)*
Hazardous air pollutants—Continued	
Policy and guidance documents	http://www.epa.gov/ttn/oarpg/amend.html
Recent regulatory developments	http://members.aol.com/CleenAir/news/newshaps.htm
Regulations	http://www.epa.gov/ttn/uatw/eparules.html
Unified Air Toxics Web site	http://www.epa.gov/ttn/uatw
Web links	http://www.4cleanair.org/toxics.html
Hazardous chemical database	http://ull.chemistry.uakron.edu/erd/sources.html
Health Effects Notebook for Hazardous Air Pollutants	http://www.epa.gov/ttn/uatw/hapindex.html
Industrial Combustion Coordinated Rulemaking	http://www.epa.gov/ttn/iccr/home.html
Industry and Trade Associations, Web links	http://members.aol.com/CleenAir/apc/indus.htm
	http://www.neis.com/associations/html
INFOSEEK search engine	http://www.infoseek.com
INFOTERRA/USA	http://www.epa.gov/earlink1/INFOTERRA/index.html
Inspections	
Chromium electroplating	http://www.epa.gov/ttn/uatw/csigc.html
Dry cleaners	http://www.epa.gov/ttn/uatw/dcsigc.html
International environmental organizations	
Consortium for International Earth Science Information Network	http://www.ciesin.org
INFOTERRA	http://www.epa.gov/earlink1/INFOTERRA/index.html
Web links	http://www.4cleanair.org/international.html
	http://www.epa.gov/ttn/chief/relsites.html#foreign
International Organization for Standardization	http://www.iso.ch
International ozone protection policy	http://www.epa.gov/docs/ozone/intpol/intpol.html
Internet resources for the environmental scientist	http://www.imt.net/~dcouncil/env.html
ISO 14000	
American National Standards Institute	http://www.ansi.org
International Organization for Standardization	http://www.iso.ch
LEXIS-NEXIS Environmental Law Library[2]	http://www.lexis-nexis.com/lncc/sources/libcont/envirn.html
Local agencies, Web links	http://www.4cleanair.org/local.html
Local Government Environmental Assistance Network	http://www.lgean.org
Locating and estimating air toxic emission documents	http://www.epa.gov/ttn/chief/ap42etc.html#LE
Looksmart search engine	http://www.looksmart.com
Lycos search engine	http://www.lycos.com

TABLE 11-1
CAA Internet Directory[1]—Continued

Site name/keyword	*Address (URL)*
MACT standards	
Pharmaceutical industry	http://206.4.171.30/rules/pharmact/index.cfm
	http://www.epa.gov/ttn/uatw/pharma/pharmpg.html
Schedules	http://www.epa.gov/ttn/uatw/2_4yrstds.html
	http://www.epa.gov/ttn/uatw/7_10yrstds.html
Material safety data sheets	
ChemFinder	http://chemfinder.camsoft.com
Hazardous chemical database	http://ull.chemistry.uakron.edu/erd/sources.html
Metal Finishing Resource Center	http://www.nmfrc.org/nmfrchom.cfm
Meteorological data	http://www.epa.gov/ttn/scram
Meteorology, training course	http://www.shodor.org/metweb/index.html
Methyl bromide	http://www.epa.gov/docs/ozone/mbr/mbrqa.html
Model permits	http://www.epa.gov/ttncaaa1/t5/meta/m2510.html
	http://www.epa.gov/ttncaaa1/t5/meta/m989.html
Monitoring	
Continuous emission monitoring	http://www.dep.state.pa.us/dep/deputate/airwaste/aq/cemspage/cemshome.htm
Aerometric Information Retrieval System	http://www.epa.gov/airs/airs.html
Ambient Monitoring Information Center	http://www.epa.gov/ttn/amtic
Ambient monitoring regulations	http://www.epa.gov/ttn/amtic/codefed.html
Emission Measurement Center	http://www.epa.gov/ttn/emc
Enhanced ozone monitoring	http://www.epa.gov/oar/oaqps/pams
EPA contacts, ambient monitoring	http://www.epa.gov/ttn/amtic/contacts.html
Montreal Protocol	http://www.ciesin.org/TG/PI/POLICY/montpro.html
NAAQS	
List of	http://www.epa.gov/airs/criteria.html
Policy and guidance documents	http://www.epa.gov/ttn/oarpg/t1main.html
Recent regulatory developments	http://members.aol.com/CleenAir/news/newsnaaq.htm
Summaries of NAAQS	http://www.epa.gov/oar/oaqps/greenbk
National Center for Environmental Assessment	http://www.epa.gov/ncea
National Center for Environmental Publications and Information	http://www.epa.gov/ncepihom/index.html
National emissions inventory data	http://www.epa.gov/ttn/chief/ei_data.html#psd
National emissions trends (1900–1996)	http://www.epa.gov/oar/emtrnd
National Environmental Laboratory Accreditation Conference	http://www.epa.gov/ttn/nelac
National Environmental Publications Information	http://www.epa.gov/cincl
National Technical Information Service	http://www.fedworld.gov/ntis/ntishome.html

TABLE 11-1
CAA Internet Directory[1]—Continued

Site name/keyword	*Address (URL)*
National Toxicology Program	http://ntp-server.niehs.nih.gov
New source review	http://www.epa.gov/ttn/nsr
Policy and guidance database	http://www.epa.gov/region07/programs/artd/air/nsr/nsrpg.htm
Nonattainment areas, lists of	http://www.epa.gov/oar/oaqps/greenbk
Office of Air and Radiation	
People locator	http://www.epa.gov/oar/oarpeopl.html
Policy and guidance information	http://www.epa.gov/ttn/oarpg
Search engine	http://www.epa.gov/oar/oarsrch.html
Office of Air Quality Planning and Standards	http://www.epa.gov/oar/oaqps
OECA	http://www.epa.gov/oeca
Office of Management and Budget, regulations under review	http://library.whitehouse.gov/omb/OMBREGS.HTM
Operating permits	
EPA operating permits Web page	http://www.epa.gov/oar/oaqps/permits
Model permits	http://www.epa.gov/ttncaaa1/t5/meta/m2510.html
	http://www.epa.gov/ttncaaa1/t5/meta/m989.html
Policy and guidance database	http://www.epa.gov/region07/programs/artd/air/title5/titlevhp.htm#Guidance
Policy and guidance documents	http://www.epa.gov/ttn/oarpg/amend.html
Recent regulatory developments	http://members.aol.com/CleenAir/news/newsti5.htm
Ozone Transport Assessment Group	http://www.epa.gov/ttn/otag
Ozone	
Enhanced ozone monitoring	http://www.epa.gov/oar/oaqps/pams
Ground-level ozone	http://www.epa.gov/docs/region01/eco/ozone/hold/index.html
Particulate matter, and regional haze implementation	http://www.epa.gov/ttn/faca
Pace Virtual Environmental Law Library	http://joshua.law.pace.edu/env/vell6.html
Paint and Coatings Resource Center	http://www.paintcenter.org
Particulate matter, Web links	http://www.4cleanair.org/particulate.html
Permits, Web links	http://www.4cleanair.org/permits.html
Pharmaceutical MACT	
Compliance assistance expert system	http://206.4.171.30/rules/pharmact/index.cfm
EPA Web site	http://www.epa.gov/ttn/uatw/pharma/pharmpg.html
Photochemical Assessment Monitoring Stations	http://www.epa.gov/oar/oaqps/pams
PKZIP/PKUNZIP	http://www.epa.gov/ttn/tools.html
Pollution Engineering Online	http://www.manufacturing.net/magazine/library/ref.htm
Pollution prevention	
ENVIROENE	http://es.epa.gov
Solvent alternatives	http://es.epa.gov/ssds/ssds.html
Web links	http://www.4cleanair.org/pp_links.html
Prevention of significant deterioration, policy and guidance database	http://www.epa.gov/region07/programs/artd/air/nsr/nsrpg.htm

©Elsevier Science Inc. 0-444-10032-6/98/$19.00+$5.50

TABLE 11-1
CAA Internet Directory[1]—Continued

Site name/keyword	*Address (URL)*
Public forums, TTN Web sites	http://www.epa.gov/ttn/forum.html
RACT/BACT/LAER Clearinghouse	http://mapsweb.rtpnc.epa.gov/RBLCWeb/bl02.htm
Real-time air pollution data	http://www.epa.gov/airnow
Regional air pollution control initiatives	http://members.aol.com/CleenAir/apc/fed2.htm#regional
Regulations	
Elsevier Science Federal Compliance Alert	http://elsevier.enviroinfo.com/frame3.shtml
Office of Air and Radiation	http://www.epa.gov/ttn/oarpg
Risk assessment	
Glossary of terms	http://www.epa.gov/ttn/uatw/hapsec1.html
Health effects	http://www.epa.gov/ttn/uatw/hapindex.html
National Center for Environmental Assessment	http://www.epa.gov/ncea
National Toxicology Program	http://ntp-server.niehs.nih.gov
Risk management plans	http://members.aol.com/cleanair4/docs/rmp-112.htm#GUID http://response.restoration.noaa.gov/cameo/dr_aloha/RMPtools http://response.restoration.noaa.gov/chemaids/rmp
Rutgers Air Pollution Training Center	http://www.eohsi.rutgers.edu/cet/airpoll/air.html
Search engines, EPA information	
Envirofacts	http://www.epa.gov/enviro
EPA People Locator (all of EPA)	http://www.epa.gov:6706/natlocdcd/owa/locator.home
EPA People Locator (OAR only)	http://www.epa.gov/oar/oarpeopl.html
EPA Web site	http://www.epa.gov/epahome/search.html
Government Information Locator Service	http://www.epa.gov.earth100/index.html
Office of Air and Radiation	http://www.epa.gov/oar/oarsrch.html
Online Library System	http://www.epa.gov/natlibra/ols.htm
TTN Web site	http://www.epa.gov/ttn/search.html
Web links	http://members.aol.com/CleenAir/apc/fed3.htm#AIRDBASE
Search engines, general	
Alta Vista	http://www.altavista.digital.com
Excite	http://www.excite.com
Infoseek	http://www.infoseek.com
Looksmart	http://www.looksmart.com
Lycos	http://www.lycos.com
WebCrawler	http://www.webcrawler.com
Yahoo	http://www.yahoo.com
Search engines, government information	http://www.epa.gov/earth100/index.html
Small business assistance program	http://www.epa.gov/ttn/sbap
Smog	http://www.4cleanair.org/smog.html
Solvents	
Carburetor cleaning	http://www.epa.gov/ttn/uatw/cardclen.pdf
Solvent cleaning alternatives	http://www.epa.gov/docs.ozone/title6/snap/solvents.html

TABLE 11-1
CAA Internet Directory[1]—Continued

Site name/keyword	*Address (URL)*
Source Classification Codes (SCC)	http://www.epa.gov/ttn/chief/scccodes.html
Source testing	http://www.dep.state.pa.us/dep/deputate/airwaste/aq/source/sts.htm
STAPPA/ALAPCO home page	http://www.4cleanair.org
State agency Web links	http://members.aol.com/CleenAir/apc/state.htm
	http://www.4cleanair.org/states.html
	http://www.envirobiz.com/govinfo/ustat.htm
	http://www.epa.gov/ttn/chief/relsites.html#state
	http://www.law.vill.edu/State-Agency/index.html
	http://www.nasire.org/ss/index.html
State courts	http://www.caprep.com/stcourt.htm
State laws	http://lawlib.wuacc.edu/washlaw/uslaw/statelaw.html
Stationary sources, enforcement policies, and guidance	http://es.epa.gov/oeca/ore/aed/comp/bcomp
Stratospheric ozone	
Enforcement actions	http://www.epa.gov/docs/ozone/enforce/enforce.html
EPA's Web site	http://www.epa.gov/docs/ozone
International ozone protection policy	http://www.epa.gov/docs/ozone/intpol/intpol.html
Methyl bromide	http://www.epa.gov/docs/ozone/mbr/mbrqa.html
Montreal Protocol	http://www.ciesin.org/TG/PI/POLICY/montpro.html
Policy and guidance documents	http://www.epa.gov/ttn/oarpg/amend.html
Regulations	http://www.epa.gov/docs/ozone/title6/usregs.html
SNAP program	http://www.epa.gov/docs/ozone/title6/snap/snap.html
Solvent cleaning alternatives	http://www.epa.gov/docs/ozone/title6/snap/solvents.html
Web links	http://www.4cleanair.org/ozone.html
Support Center for Regulatory Air Models	http://www.epa.gov/ttn/scram
Test methods	http://www.epa.gov/epahome/index
THOMAS federal legislative information	http://thomas.loc.gov/home/thomas2.html
Toxics release inventories	
Automated Form R software	http://www.epa.gov/opptintr/afr
Information resources	http://www.epa.gov/opptintr/tri
List of EPCRA chemicals	http://www.epa.gov/swercepp/gen-pubs.html
Sector-specific guidance	http://www.epa.gov/opptintr/tri/industry.htm
Training courses	
Air & Waste Management Association	http://www.awma.org/txtindex.htm
Air Pollution Distance Learning Network	http://www.epa.gov/oar/oaq_apti.html
Ambient Monitoring Technology Information Center	http://www.epa.gov/ttn/amtic
Education and Outreach Group	http://www.epa.gov/oar/oaqps/eog
Online air quality meteorology course	http://www.shodor.org/metweb/index.html
Rutgers Air Pollution Training Center	http://www.eohsi.rutgers.edu/cet/airpoll/air.html
Schedule of upcoming air pollution training courses	http://www.4cleanair.org/schedule.html

TABLE 11-1
CAA Internet Directory[1]—Continued

Site name/keyword	*Address (URL)*
Transportation Environmental Resource Center	http://www.transource.org
TTN Web site	http://www.epa.gov/ttn
Directory	http://www.epa.gov/ttn/direct.html
Home page	http://www.epa.gov/ttn
Search engine	http://www.epa.gov/ttn/search.html
U.S. Code	http://www.access.gpo.gov/congress/cong013.html
U.S. Court of Appeals Web sites	
DC Circuit	http://www.ll.georgetown.edu/Fed-Ct/cadc.html
First Circuit	http://www.law.emory.edu/1circuit
Second Circuit	http://www.TouroLaw.edu/2ndCircuit
Third Circuit	http://www.law.vill.edu/Fed-Ct/ca03.html
Fourth Circuit	http://www.law.emory.edu/4circuit
Fifth Circuit	http://www.ca5.uscourts.gov
Sixth Circuit	http://www.law.emory.edu/6circuit
Seventh Circuit	http://www.kentlaw.edu/7circuit
Eighth Circuit	http://www.wulaw.wustl.edu/8th.cir
Ninth Circuit	http://www.findlaw.com/casecode/courts/9th.html
Tenth Circuit	http://www.law.emory.edu/10circuit
Eleventh Circuit	http://www.law.emory.edu/11circuit
U.S. government bookstores	http://www.access.gpo.gov/su_docs/sale/abkst001.html
U.S. House of Representative Internet Law Library	http://law.house.gov
U.S. Supreme Court	http://supct.law.cornell.edu/supct/directory/united_states.html
Unified Air Toxics Website	http://www.epa.gov/ttn/uatw/
Unit conversion engines	http://chemserv.bc.edu/Data/chemicaldata.html
Unit conversion engines	http://www.webcom.com/legacysy/convert2/convert2.html
Washburn University School of Law Web site	http://lawlib.wuacc.edu/washlaw/washlaw
WebCrawler search engine	http://www.webcrawler.com
Woodstoves	http://es.epa.gov/oeca/metd/stove.html
WWW Virtual Library: Environment	http://earthsystems.org/Environment.html
Yahoo search engine	http://www.yahoo.com

ALAPCO = Association of Local Air Pollution Control Officials; BACT = best available control technology; EPCRA = Emergency Planning and Community Right-to-Know Act; LAER = lowest achievable emission rate; MACT = maximum achievable control technology; NAAQS = national ambient air quality standards; NO_x = nitrogen oxides; OAR = Office of Air & Radiation; OECA = Office of Enforcement and Compliance Assurance; RACT = reasonably available control technology; SNAP = significant new alternatives policy; SO_2 = sulfur dioxide; STAPPA = State and Territorial Air Pollution Program Administrators; TTN = Technology Transfer Network; URL = uniform resource locator.

[1]This directory includes select web sites; URLs are current as of October 1, 1998.

[2]Subscription fee required.

Source: Elsevier Science Inc.

12 Professional Toolbox

After a few years in the field, most environmental professionals have accumulated a horde of useful data, information, and facts. This information can include conversion factors, formulas, or "rules-of-thumb." All too often, the pages become blurred from repeated copying or are stashed in a manila folder or three-ring binder.

With this new chapter of *The Air Pollution Consultant Quick Reference Guide*, we provide a distillation of those oft-needed reference facts. We've chosen to title this chapter "Professional Toolbox" in the hope that its contents will be considered as vital as any craftsman's tools.

The toolbox consists of the following:

1. *General chemical information*—This section provides chemical formula, structural arrangement, and general information (including CERCLA reportable quantities [RQs]) for 50 of the most common organic air pollutants (pages 12.2–12.11). [Note: An entry of NA means the information is not applicable.]
2. *Vapor pressure charts*—These handy, easy-to-read charts graphically illustrate the relationship between temperature and air pressure for some of the organic chemicals listed in the previous section (Figures 12-1 through 12-6; pages 12.12–12.14).
3. *Combustibility data*—This table presents lower and upper explosive limits and the closed-cup flash point for selected pure gases and vapors in air (Table 12-1; pages 12.15–12.16).
4. *Auto-ignition temperatures*—The auto-ignition temperatures of some common organic compounds are presented in Table 12-2 (page 12.17).
5. *Natural gas*—This table presents some physical and chemical constants of natural gas constituents (Table 12-3; page 12.18).
6. *Particle size*—The relationship between particle size and abatement equipment is graphically presented in Figure 12-7 (page 12.19).
7. *Emissions and control technologies*—Some useful data pertaining to the nature of emission streams and various control technologies are presented (Table 12-4: page 12.20 Figure 12-8; page 12.21).
8. *Composition of air*—Both major and trace constituents of air are presented (Table 12-5; page 12.21).
9. *EPA test methods*—Promulgated, proposed, approved alternative, conditional, and preliminary test methods and procedures for emission testing and monitoring are identified (Table 12-6; pages 12.22–12.24).
10. *Conversion factors*—A collection of useful constants and conversion factors is presented in (Table 12-7; pages 12.25–12.27).

References

Bruner, C.R. *Incineration Systems: Selection and Design*. Incinerator Consultants, Inc., Reston, Virginia, 1984.

"Physical/Chemical Data Compendium for Hazardous Air Pollutants," *The Air Pollution Consultant*. Elsevier Science Inc., New York, New York, Volume 2, Issue 1, pp 4.1–4.38 (January/February 1992).

Heskith, H.E. *Air Pollution Control: Traditional and Hazardous Pollutants*. Technomic Publishing Company, Inc., Lancaster, Pennsylvania, 1991.

Kirk, R.E. et al. Concise Encyclopedia of Chemical Technology. John Wiley & Sons, Inc., New York, New York, 1985.

Manahan, S.E. Environmental Science and Technology. CRC Press LLC, Boca Raton, Florida, 1997.

Control Technologies for Hazardous Air Pollutants, EPA/625/6-91/014. U.S. Government Printing Office, Washington, DC, June 1991. APC

Acetaldehyde

Formula: C_2H_4O

Structure:

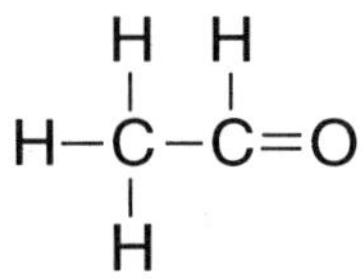

CAS number: 75-07-0

Synonyms: Acetic aldehyde, ethanal, ethyl aldehyde

Molecular weight: 44.05

CERCLA RQ: 1,000 pounds

Saturated vapor pressure @ 70°F: 15.100 psia

Acetonitrile

Formula: C_2H_3N

Structure:

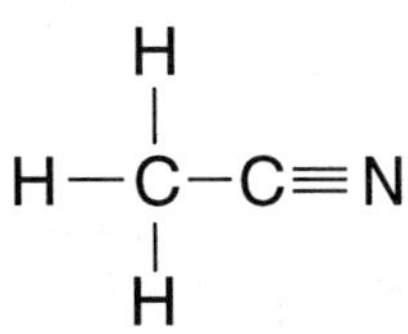

CAS number: 75-05-8

Synonyms: Cyanomethane, ethanenitrile, ethyl nitrile, methanecarbonitrile, methyl cyanide

Molecular weight: 41.05

CERCLA RQ: 5,000 pounds

Saturated vapor pressure @ 70°F: 1.383 psia

Acrylonitrile

Formula: C_3H_3N

Structure:

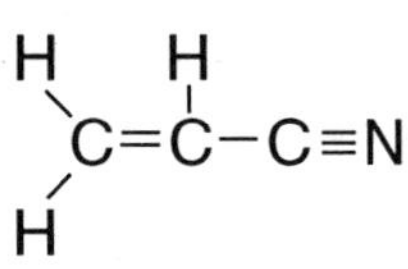

CAS number: 107-13-1

Synonyms: Acrylon, carbacryl, cyanoethylene, fumigrain, propenenitrile, vinyl cyanide

Molecular weight: 53.60

CERCLA RQ: 100 pounds

Saturated vapor pressure @ 70°F: 1.707 psia

Benzene

Formula: C_6H_6

Structure:

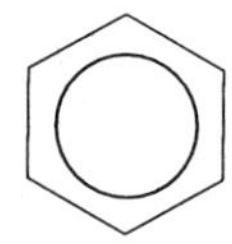

CAS number: 71-43-2

Synonyms: Benzol, coal naphtha, cyclohexatriene, phene, phenyl hydride, pyrobenzol

Molecular weight: 78.11

CERCLA RQ: 10 pounds

Saturated vapor pressure @ 70°F: 1.535 psia

Biphenyl

Formula: C_6H_5

Structure:

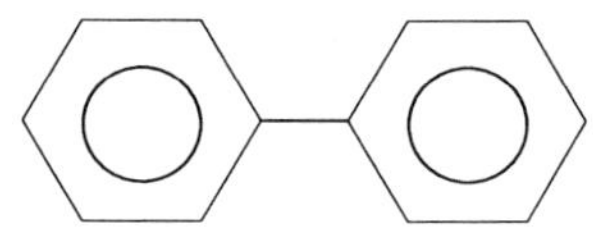

CAS number: 92-52-4

Synonyms: Bibenzene; 1,1′-biphenyl; diphenyl; 1,1′-diphenyl; phenylbenzene

Molecular weight: 154.22

CERCLA RQ: 100 pounds

Saturated vapor pressure @ 70°F: NA

Bis(2-ethylhexyl)phthalate

Formula: $C_{24}H_{38}O_4$

Structure:

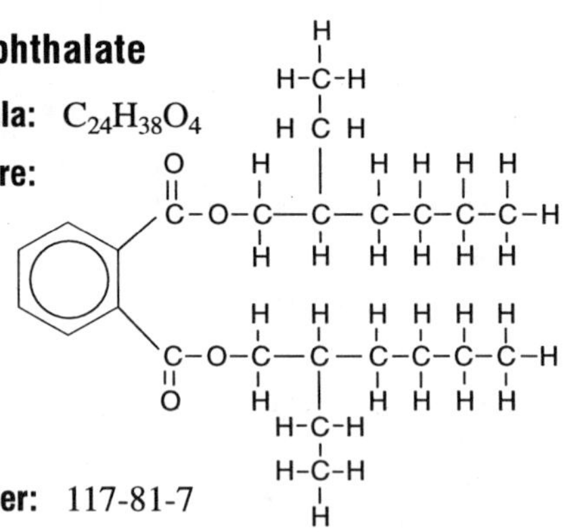

CAS number: 117-81-7

Synonyms: 1,2-Benzenedicarboxylic acid, bis(2-ethylhexyl)ester; DEHP; di-(2-ethylhexyl) phthalate; dioctyl phthalate; DOP; phthalic acid

Molecular weight: 390.62

CERCLA RQ: 100 pounds

Saturated vapor pressure @ 70°F: NA

1,3-Butadiene

Formula: C_4H_6

Structure:

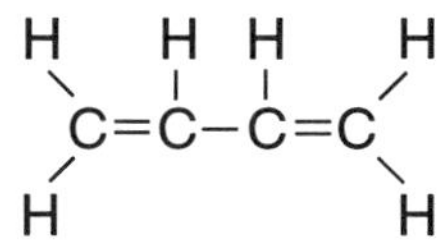

CAS number: 106-99-0

Synonym: Butadiene

Molecular weight: 54.10

CERCLA RQ: 10 pounds

Saturated vapor pressure @ 70°F: 36.050 psia

Carbon disulfide

Formula: CS_2

Structure: S=C=S

CAS number: 75-15-01

Synonyms: Carbon bisulfide, carbon sulfide, dithiocarbonic anhydride

Molecular weight: 76.13

CERCLA RQ: 100 pounds

Saturated vapor pressure @ 70°F: 5.990 psia

Carbon tetrachloride

Formula: CCl_4

Structure:

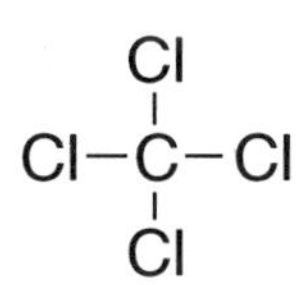

CAS number: 56-23-5

Synonyms: Benzinoform, carbon chloride, perchloromethane, R-10, tetrachloromethane, tetraform

Molecular weight: 153.83

CERCLA RQ: 10 pounds

Saturated vapor pressure @ 70°F: 1.874 psia

Carbonyl sulfide

Formula: COS

Structure: O=C=S

CAS number: 463-58-1

Synonym: Carbon oxide sulfide

Molecular weight: 60.07

CERCLA RQ: 100 pounds

Saturated vapor pressure @ 70°F: NA

Chlorine

Formula: Cl_2

Structure: Cl–Cl

CAS number: 7782-50-5

Synonym: Molecular chlorine

Molecular weight: 70.91

CERCLA RQ: 10 pounds

Saturated vapor pressure @ 70°F: 94.300 psia

Chlorobenzene

Formula: C_6H_5Cl

Structure:

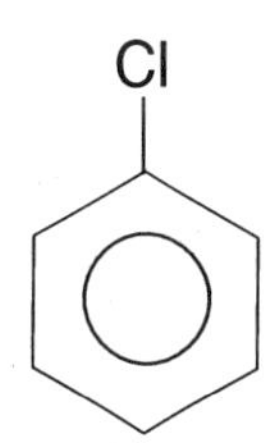

CAS number: 108-90-7

Synonyms: MCB, monochlorobenzene, phenyl chloride

Molecular weight: 112.56

CERCLA RQ: 100 pounds

Saturated vapor pressure @ 70°F: 0.195 psia

Chloroform

Formula: $CHCl_3$

Structure:

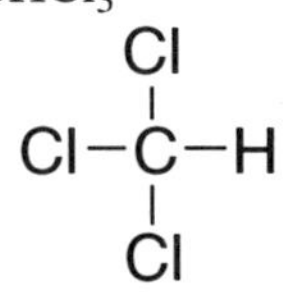

CAS number: 67-66-3

Synonyms: Formyl trichloride, methane trichloride, methenyl chloride, R-20, trichloroform, trichloromethane

Molecular weight: 119.39

CERCLA RQ: 10 pounds

Saturated vapor pressure @ 70°F: 3.229 psia

Chloroprene

Formula: C_4H_5Cl

Structure:

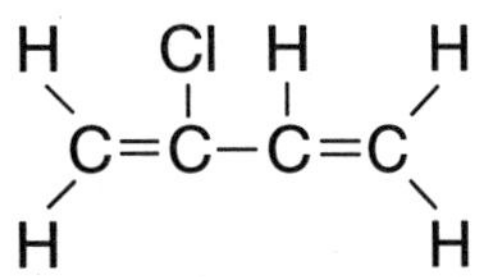

CAS number: 126-99-8

Synonyms: 2-Chloro-1,3-butadiene; 2-chlorobutadiene; neoprene

Molecular weight: 88.54

CERCLA RQ: 100 pounds

Saturated vapor pressure @ 70°F: NA

Cresols (mixed isomers)

Formula: C_7H_8O

Structure:

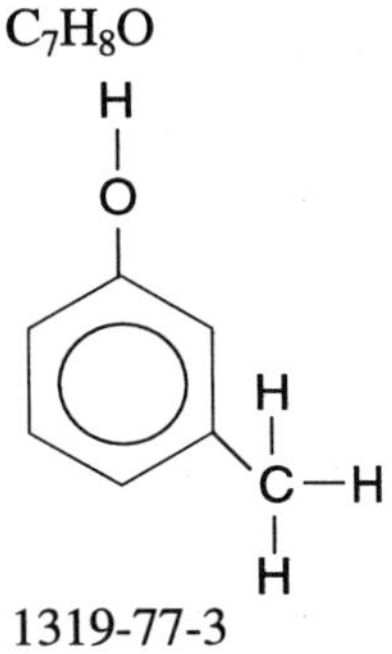

CAS number: 1319-77-3

Synonyms: Cresylic acid, methyl phenol, tricresol

Molecular weight: 108.15

CERCLA RQ: 100 pounds

Saturated vapor pressure @ 70°F: 0.110 psia

Cumene

Formula: C_9H_{12}

Structure:

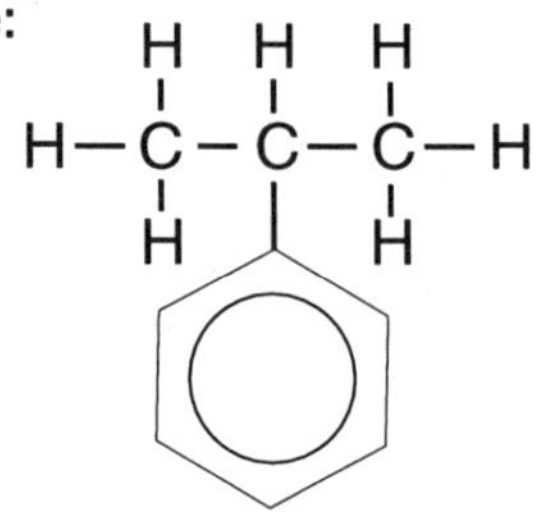

CAS number: 98-82-8

Synonyms: Cumol, isopropylbenzene, 1-(methylethyl)-benzene, 2-phenylpropane

Molecular weight: 120.19

CERCLA RQ: 5,000 pounds

Saturated vapor pressure @ 70°F: 0.272 psia

1,4-Dichlorobenzene

Formula: $C_6H_4Cl_2$

Structure:

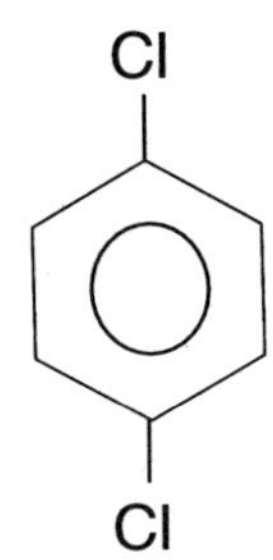

CAS number: 106-46-7

Synonyms: *p*-Chlorophenyl chloride, *p*-dichlorobenzene

Molecular weight: 147.01

CERCLA RQ: 100 pounds

Saturated vapor pressure @ 70°F: 0.407 psia

Ethyl benzene

Formula: C_8H_{10}

Structure:

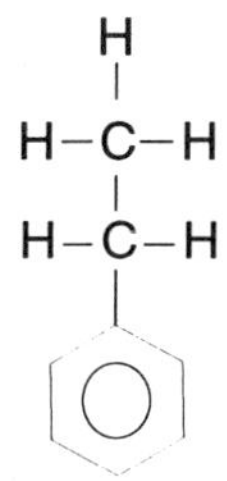

CAS number: 100-41-4

Synonyms: EB, ethylbenzene, ethylbenzol, phenylethane

Molecular weight: 106.17

CERCLA RQ: 1,000 pounds

Saturated vapor pressure @ 70°F: 0.160 psia

Ethyl chloride

Formula: C_2H_5Cl

Structure:

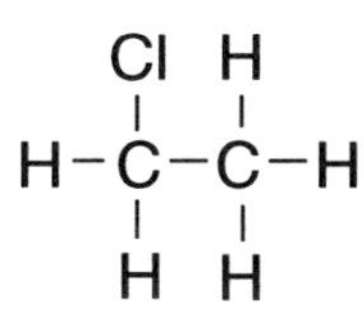

CAS number: 75-00-3

Synonyms: Chlorene, chlorethyl, chloroethane, chloryl, ether muriatic, hydrochloric ether, monochloroethane

Molecular weight: 64.52

CERCLA RQ: 100 pounds

Saturated vapor pressure @ 70°F: 20.260 psia

Ethylene dichloride

Formula: $C_2H_4Cl_2$

Structure:

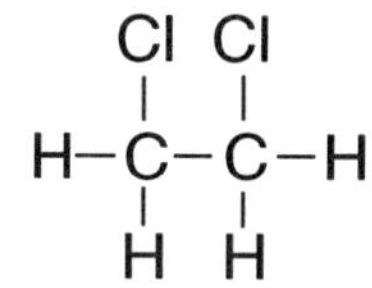

CAS number: 107-06-2

Synonyms: 1,2-Bichloroethane; brocide; 1,2-dichloroethane; *sym*-dichloroethane; EDC; ethylene chloride; glycol dichloride

Molecular weight: 98.96

CERCLA RQ: 100 pounds

Saturated vapor pressure @ 70°F: 1.246 psia

Ethylene glycol

Formula: $C_2H_6O_2$

Structure:

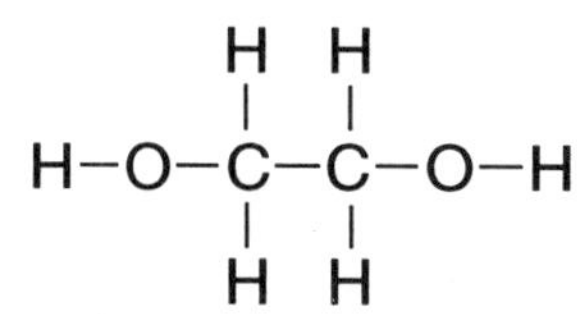

CAS number: 107-21-1

Synonyms: 1,2-Dihydroxyethane; 1,2-ethanediol; ethylene alcohol; ethylene dihydrate; glycol; glycol alcohol; 2-hydroxyethanol; monoethylene glycol

Molecular weight: 62.07

CERCLA RQ: 5,000 pounds

Saturated vapor pressure @ 70°F: 0.001 psia

Ethylene oxide

Formula: C_2H_4O

Structure:

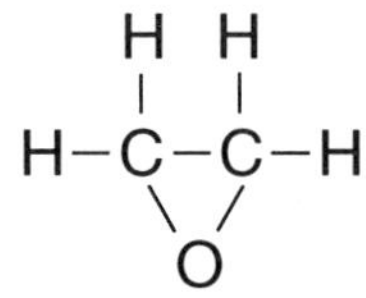

CAS number: 75-21-8

Synonyms: Dihydrooxirene; dimethylene oxide; ETO; 1,2-epoxyethane; ethene oxide; T-gas; oxacyclopropane; oxane; oxidoethane; oxirane

Molecular weight: 44.06

CERCLA RQ: 10 pounds

Saturated vapor pressure @ 70°F: 21.900 psia

Formaldehyde

Formula: CH_2O

Structure:

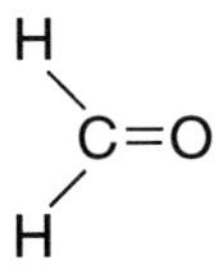

CAS number: 50-00-0

Synonyms: Formic aldehyde, formol, methanal, methyl aldehyde, methylene oxide, oxomethane, oxymethylene

Molecular weight: 30.03

CERCLA RQ: 100 pounds

Saturated vapor pressure @ 70°F: NA

Hydrochloric acid

Formula: HCl

Structure: H—Cl

CAS number: 7647-01-0

Synonyms: Chlorohydric acid, hydrochloride, hydrogen chloride, muriatic acid

Molecular weight: 36.47

CERCLA RQ: 5,000 pounds

Saturated vapor pressure @ 70°F: 60.200 psia

Hydrogen fluoride

Formula: HF

Structure: H—F

CAS number: 7664-39-3

Synonym: Hydrofluoric acid

Molecular weight: 20.01

CERCLA RQ: 100 pounds

Saturated vapor pressure @ 70°F: 15.120 psia

Methanol

Formula: CH_4O

Structure:

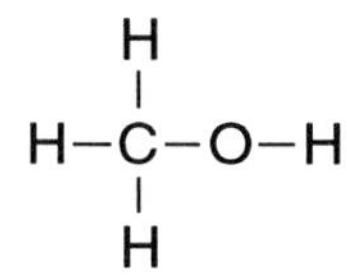

CAS number: 67-56-1

Synonyms: Carbinol, methyl alcohol, methyl hydroxide, methylol, monohydroxymethane, wood alcohol

Molecular weight: 32.04

CERCLA RQ: 5,000 pounds

Saturated vapor pressure @ 70°F: 1.930 psia

Methyl bromide

Formula: CH_3Br

Structure:

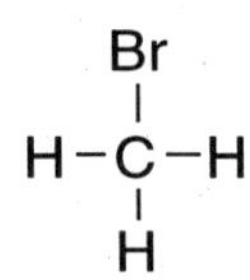

CAS number: 74-83-9

Synonyms: Bromomethane, curafume, embafume, Halon-1001, monobromomethane

Molecular weight: 94.95

CERCLA RQ: 1,000 pounds

Saturated vapor pressure @ 70°F: 16.400 psia

Methyl *tert*-butyl ether

Formula: $C_5H_{12}O$

Structure:

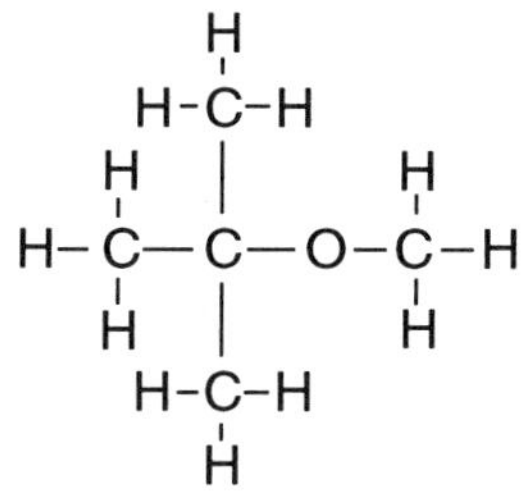

CAS number: 1634-04-4

Synonyms: Butyl methyl ether, 2-methoxy-2-methyl propane, MTBE

Molecular weight: 88.15

CERCLA RQ: 1,000 pounds

Saturated vapor pressure @ 70°F: NA

Methyl chloride

Formula: CH_3Cl

Structure:

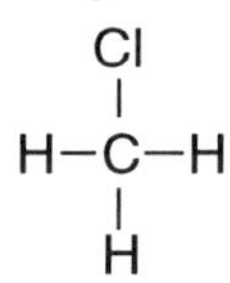

CAS number: 74-87-3

Synonyms: Chloromethane, monochloromethane, R-40

Molecular weight: 50.49

CERCLA RQ: 100 pounds

Saturated vapor pressure @ 70°F: 71.500 psia

Methyl chloroform

Formula: $C_2H_3Cl_3$

Structure:

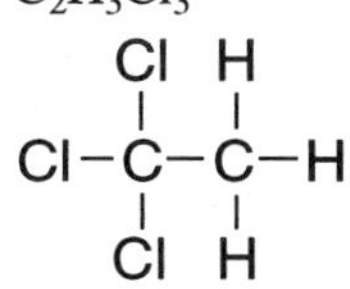

CAS number: 71-55-6

Synonyms: 1,1,1-Trichloroethane, TCA

Molecular weight: 133.40

CERCLA RQ: 1,000 pounds

Saturated vapor pressure @ 70°F: 2.099 psia

Methyl ethyl ketone

Formula: C_4H_8O

Structure:

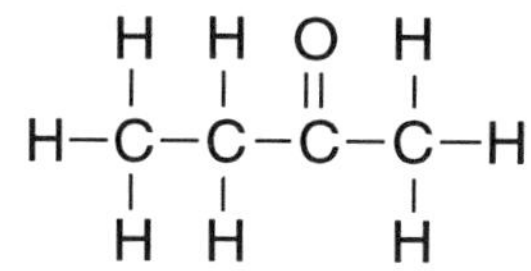

CAS number: 78-93-3

Synonyms: Butanone, 2-butanone, ethyl methyl ketone, MEK

Molecular weight: 72.11

CERCLA RQ: 5,000 pounds

Saturated vapor pressure @ 70°F: 1.461 psia

Methyl isobutyl ketone

Formula: $C_6H_{12}O$

Structure:

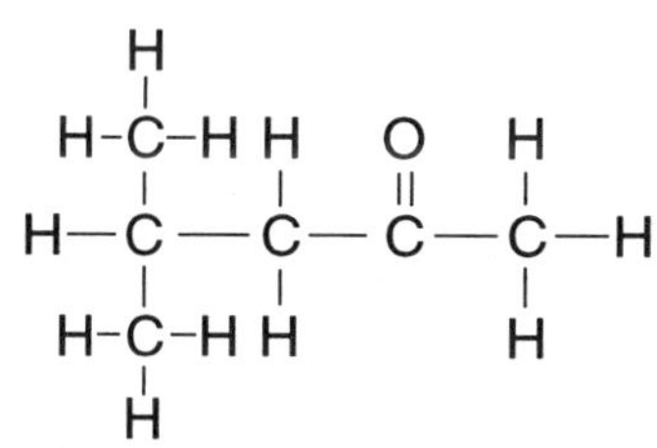

CAS number: 108-10-1

Synonyms: Hexone, isopropylacetone, isobutyl methyl ketone, 2-methyl-4-pentanone, 2-methypropyl methyl ketone, 4-methyl-2-oxopentane, MIBK, MIK

Molecular weight: 100.16

CERCLA RQ: 5,000 pounds

Saturated vapor pressure @ 70°F: 0.308 psia

Methyl methacrylate

Formula: $C_5H_8O_2$

Structure:

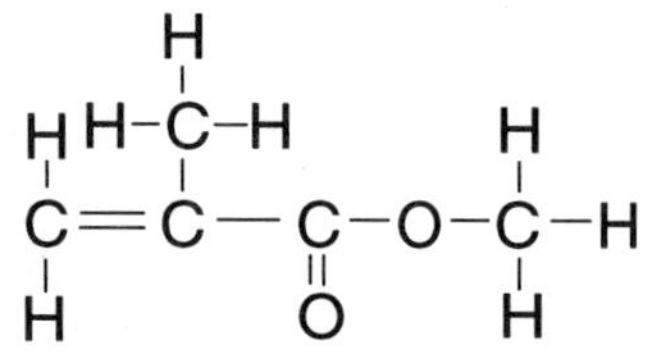

CAS number: 80-62-6

Synonyms: Methacrylic acid methyl ester, methyl 2-methyl-2-propenoate, MME,

Molecular weight: 100.12

CERCLA RQ: 1,000 pounds

Saturated vapor pressure @ 70°F: 0.635 psia

Methylene chloride

Formula: CH_2Cl_2

Structure:

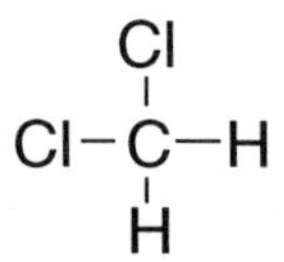

CAS number: 75-09-2

Synonyms: Dichloromethane, methane dichloride, methylene chloride, methylene dichloride, R-30

Molecular weight: 84.93

CERCLA RQ: 1,000 pounds

Saturated vapor pressure @ 70°F: 7.389 psia

Naphthalene

Formula: $C_{10}H_8$

Structure:

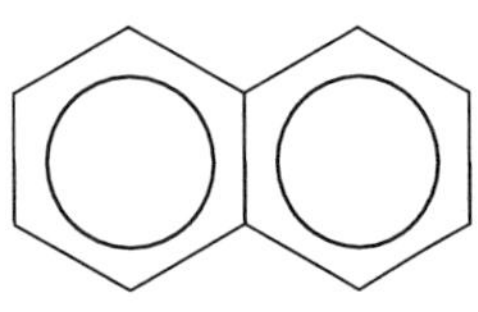

CAS number: 91-20-3

Synonyms: Moth flakes, naphthaline, naphthene, tar camphor, white tar

Molecular weight: 128.18

CERCLA RQ: 100 pounds

Saturated vapor pressure @ 70°F: NA

Phenol

Formula: C_6H_6O

Structure:

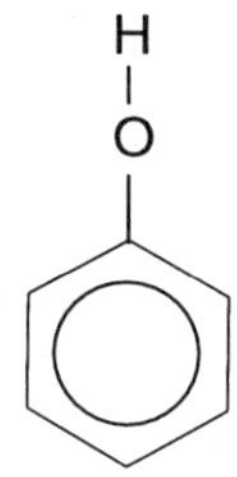

CAS number: 108-95-2

Synonyms: Benzenol, carbolic acid, hydroxybenzene, monohydroxybenzene, monophenol, oxybenzene, phenic acid, phenyl alcohol, phenyl hydroxide, phenylic acid

Molecular weight: 94.11

CERCLA RQ: 1,000 pounds

Saturated vapor pressure @ 70°F: 0.012 psia

Propionaldehyde

Formula: C_3H_6O

Structure:

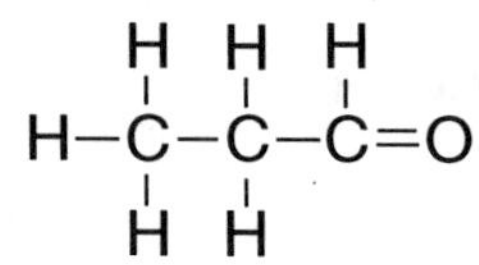

CAS number: 123-38-6

Synonyms: Methylacetaldehyde, propaldehyde, propanaldehyde, propanol, propional, propionic aldehyde, propylaldehyde

Molecular weight: 58.08

CERCLA RQ: 1,000 pounds

Saturated vapor pressure @ 70°F: 5.214 psia

Propylene dichloride

Formula: $C_3H_6Cl_2$

Structure:

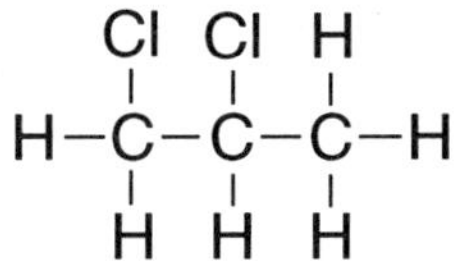

CAS number: 78-87-5

Synonyms: 1,2-Dichloropropane; propylene chloride

Molecular weight: 112.99

CERCLA RQ: 1,000 pounds

Saturated vapor pressure @ 70°F: 0.836 psia

Propylene oxide

Formula: C_3H_6O

Structure:

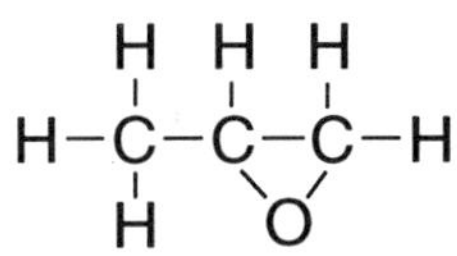

CAS number: 75-56-9

Synonyms: 1,2-Epoxypropane; methyloxirane; propene oxide; 1,2-propylene oxide

Molecular weight: 58.08

CERCLA RQ: 100 pounds

Saturated vapor pressure @ 70°F: 9.050 psia

Styrene

Formula: C_8H_8

Structure:

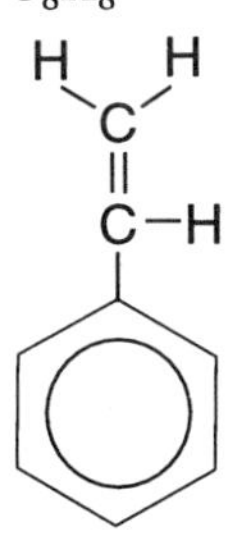

CAS number: 100-42-5

Synonyms: Cinnamene, ethenyl benzene, phenylethylene, styrol, styrolene, vinylbenzene, vinylbenzol

Molecular weight: 104.16

CERCLA RQ: 1,000 pounds

Saturated vapor pressure @ 70°F: 0.120 psia

Tetrachloroethylene

Formula: C_2Cl_4

Structure:

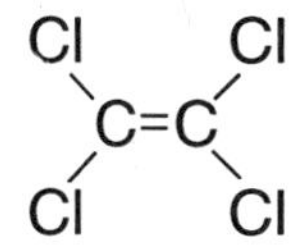

CAS number: 127-18-4

Synonyms: PCE, perchloroethylene, tetrachloroethene

Molecular weight: 165.83

CERCLA RQ: 100 pounds

Saturated vapor pressure @ 70°F: 0.318 psia

Toluene

Formula: C_7H_8

Structure:

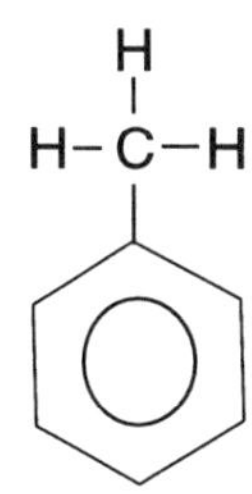

CAS number: 108-88-3

Synonyms: Methylbenzene, methylbenzol, phenylmethane, toluol

Molecular weight: 92.14

CERCLA RQ: 1,000 pounds

Saturated vapor pressure @ 70°F: 0.449 psia

1,2,4-Trichlorobenzene

Formula: $C_6H_3Cl_3$

Structure:

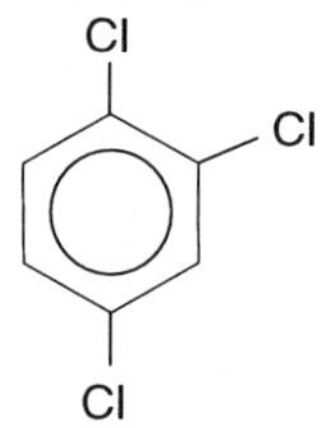

CAS number: 120-82-1

Synonyms: Trichlorobenzene; 1,2,4-trichlorobenzol 1,2,5

Molecular weight: 181.45

CERCLA RQ: 100 pounds

Saturated vapor pressure @ 70°F: NA

Trichloroethylene

Formula: C_2HCl_3

Structure:

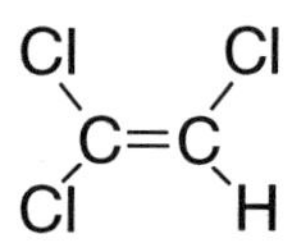

CAS number: 79-01-6

Synonyms: Ethinyl trichloride, ethylene trichloride, TCE, trichloroethene

Molecular weight: 131.19

CERCLA RQ: 100 pounds

Saturated vapor pressure @ 70°F: 1.166 psia

Vinyl acetate

Formula: $C_4H_6O_2$

Structure:

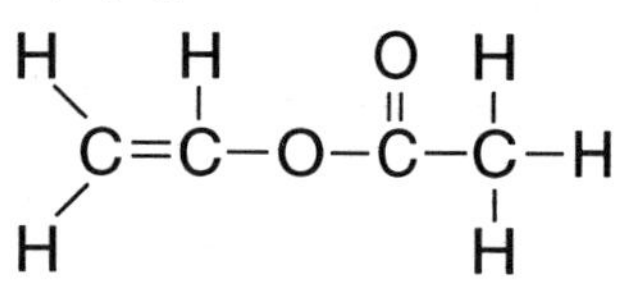

CAS number: 108-05-4

Synonyms: Acetic acid ethenyl ester, acetic acid vinyl ester, 1-acetoxyethylene, ethenyl acetate, vinyl A monomer

Molecular weight: 86.09

CERCLA RQ: 5,000 pounds

Saturated vapor pressure @ 70°F: 1.871 psia

Vinyl chloride

Formula: C_2H_3Cl

Structure:

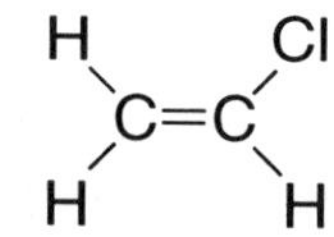

CAS number: 75-01-4

Synonyms: Chloroethene, chloroethylene, monochloroethylene, vinyl C monomer

Molecular weight: 62.50

CERCLA RQ: 1 pound

Saturated vapor pressure @ 70°F: 46.600 psia

Xylene (mixed isomers)

Formula: C_8H_{10}

Structure: **See *m*-, *o*-, and *p*-Xylene**

CAS number: 1330-20-7

Synonyms: Dimethylbenzene; xylol

Molecular weight: 106.16

CERCLA RQ: 100 pounds

Saturated vapor pressure @ 70°F: 0.130 psia

m-Xylene

Formula: C_8H_{10}

Structure:

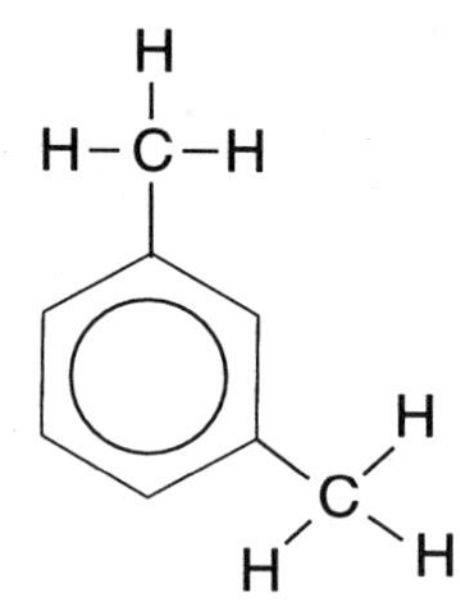

CAS number: 108-38-3

Synonyms: 1,3-Dimethylbenzene; *m*-dimethylbenzene; xylol

Molecular weight: 106.16

CERCLA RQ: 1,000 pounds

Saturated vapor pressure @ 70°F: 0.127 psia

o-Xylene

Formula: C_8H_{10}

Structure:

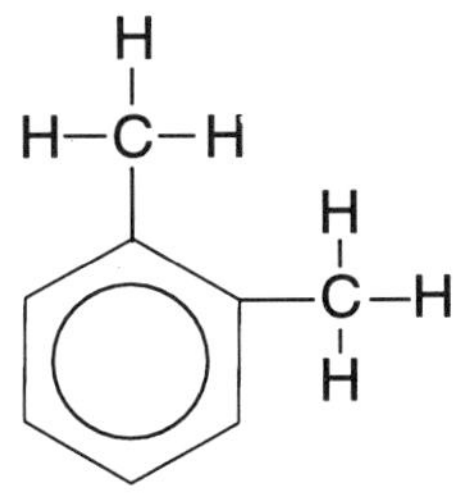

CAS number: 95-47-6

Synonyms: 1,2-Dimethylbenzene; *o*-dimethylbenzene; xylol

Molecular weight: 106.16

CERCLA RQ: 1,000 pounds

Saturated vapor pressure @ 70°F: 0.101 psia

p-Xylene

Formula: C_8H_{10}

Structure:

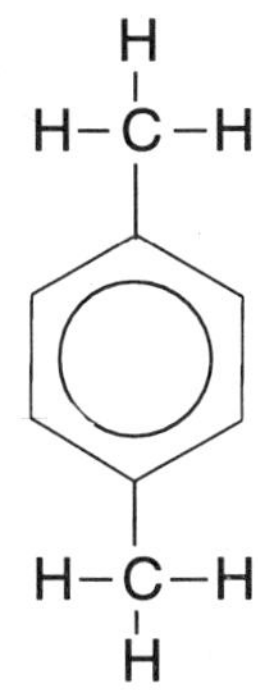

CAS number: 106-42-3

Synonyms: 1,4-Dimethylbenzene; *p*-dimethylbenzene; xylol

Molecular weight: 106.16

CERCLA RQ: 100 pounds

Saturated vapor pressure @ 70°F: 0.135 psia

FIGURE 12-1
Vapor Pressures of Selected Hazardous Air Pollutants

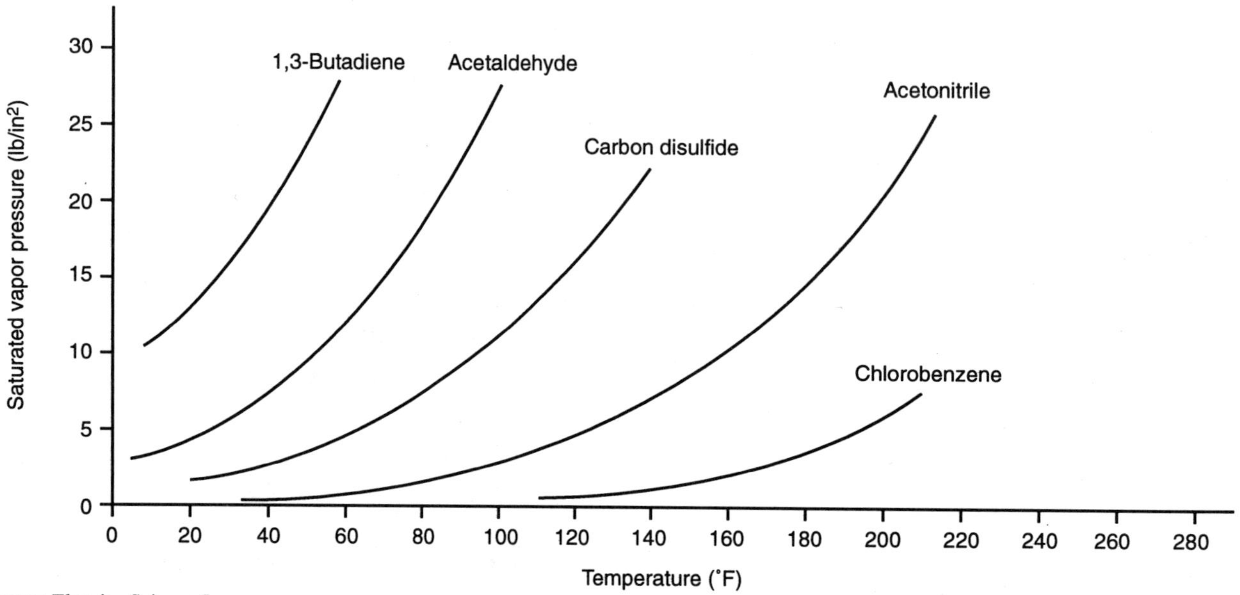

Source: Elsevier Science Inc.

FIGURE 12-2
Vapor Pressures of Selected Hazardous Air Pollutants

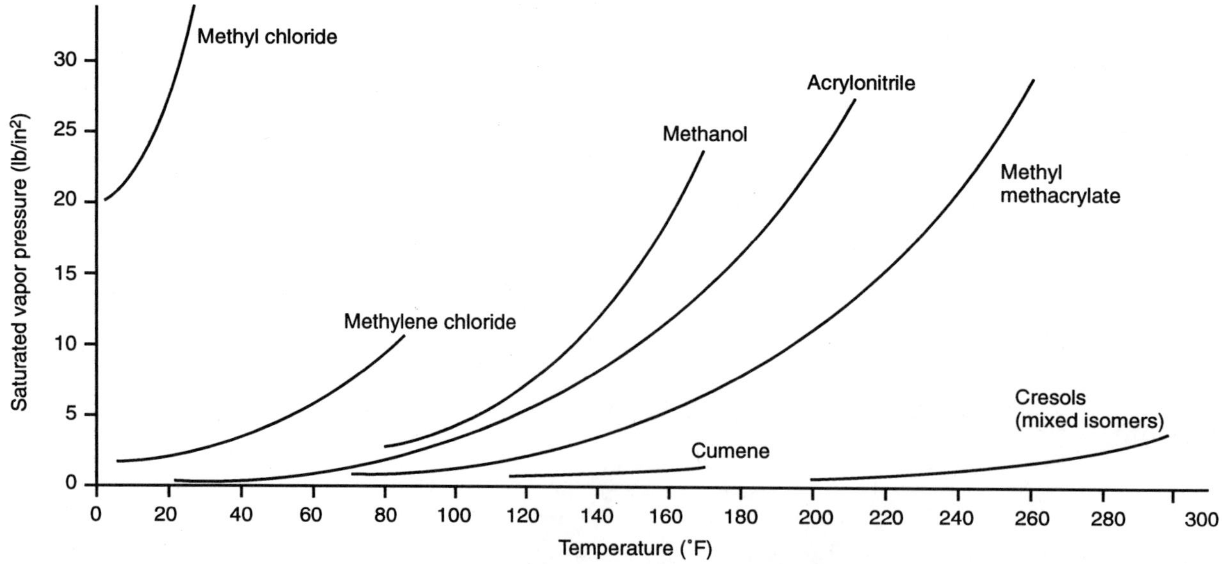

Source: Elsevier Science Inc.

FIGURE 12-3
Vapor Pressures of Selected Hazardous Air Pollutants

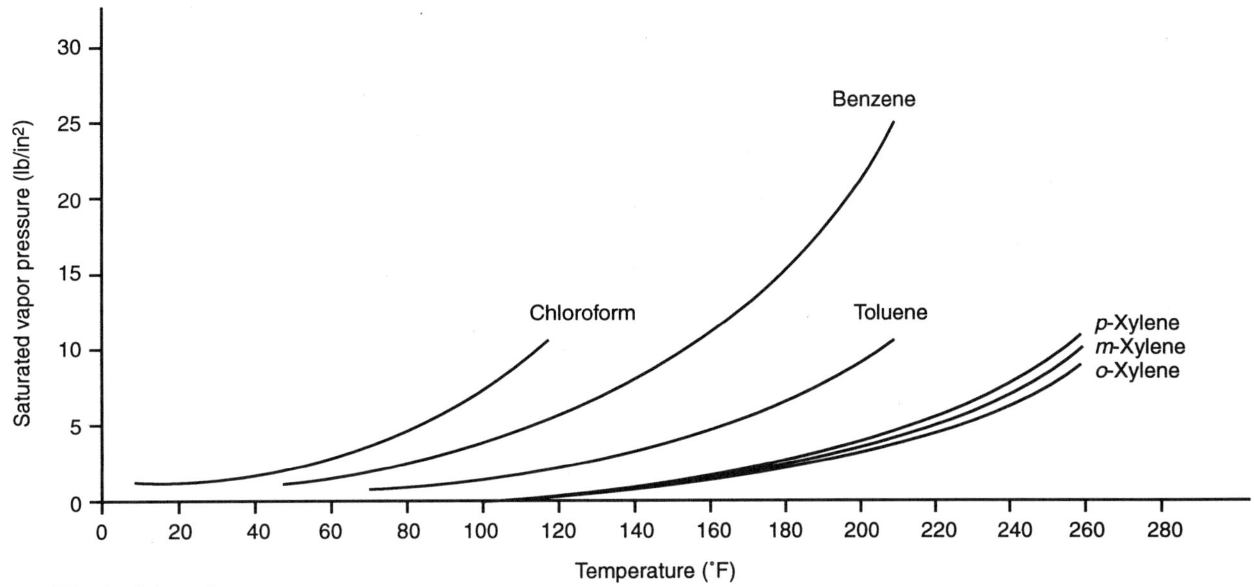

Source: Elsevier Science Inc.

FIGURE 12-4
Vapor Pressures of Selected Hazardous Air Pollutants

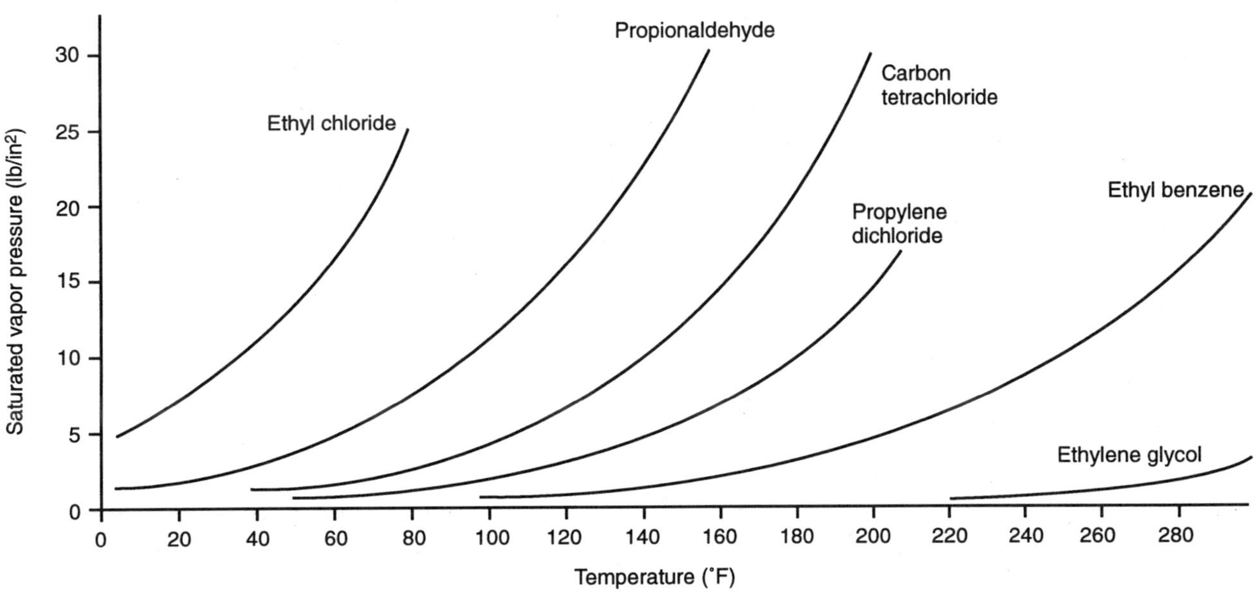

Source: Elsevier Science Inc.

FIGURE 12-5
Vapor Pressures of Selected Hazardous Air Pollutants

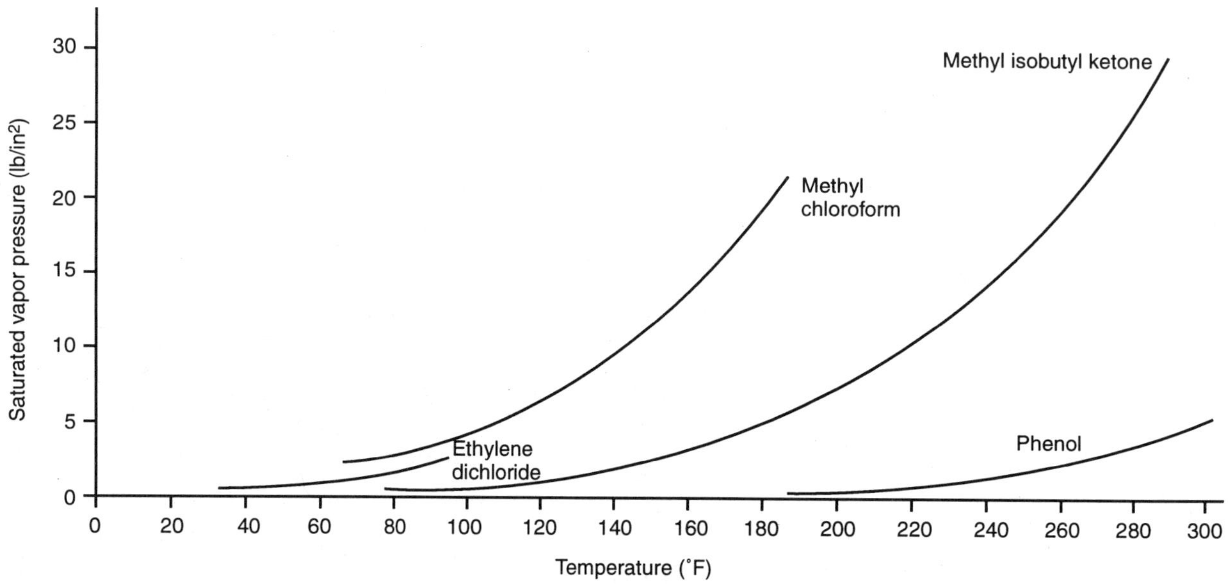

Source: Elsevier Science Inc.

FIGURE 12-6
Vapor Pressures of Selected Hazardous Air Pollutants

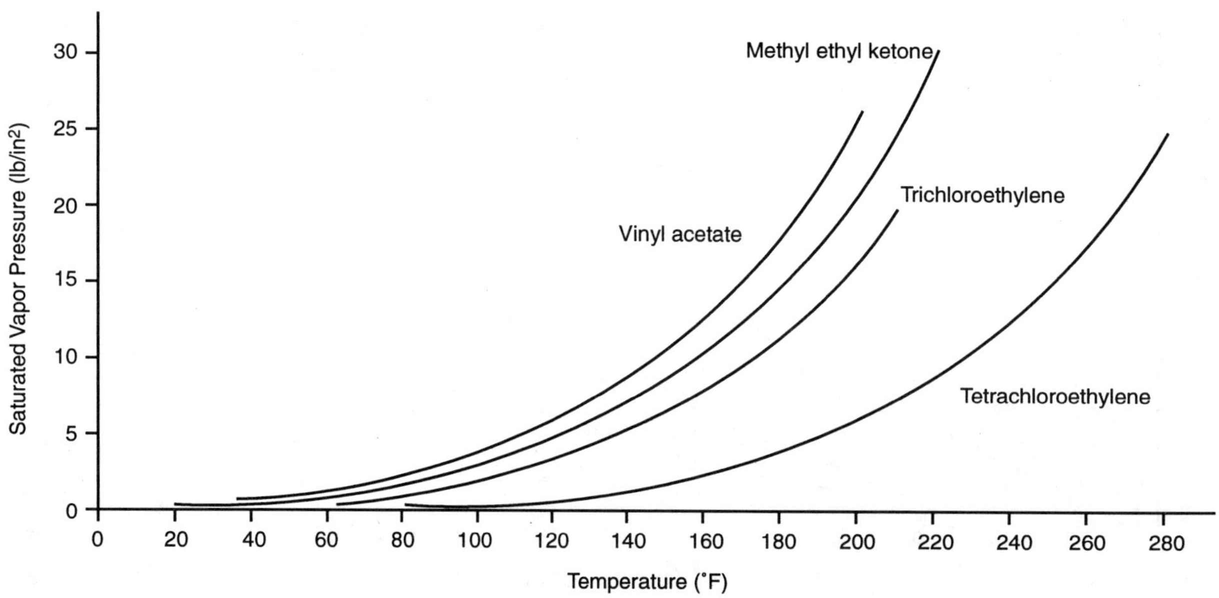

Source: Elsevier Science Inc.

TABLE 12-1
Combustibility Characteristics of Selected Pure Gases and Vapors in Air

Gas or vapor	Lower limit (% by volume)	Upper limit (% by volume)	Closed-cup flash point (°F)
Acetaldehyde	4.0	57	–17
Acetone	2.5	12.8	0
Acetylene	2.5	80	—
Allyl alcohol	2.5	—	70
Ammonia	15.5	26.6	—
Amyl acetate	1.0	7.5	77
Amylene	1.6	7.7	—
Benzene (benzol)	1.3	6.8	12
Benzlyl chloride	1.1	—	140
Butene	1.8	8.4	—
Butyl acetate	1.4	15.0	84
Butyl alcohol	1.7	—	—
Butyl cellosolve	—	—	141
Carbon disulfide	1.2	50	–22
Carbon monoxide	12.5	74.2	—
Chlorobenzene	1.3	7.1	90
Cottonseed oil	—	—	486
Cresol (*m*- or *p*-)	1.1	—	202
Crotonaldehyde	2.1	15.5	55
Cyclohexane	1.3	8.4	1
Cyclohexanone	1.1	—	111
Cyclopropane	2.4	10.5	—
Cymene	0.7	—	117
Dichlorobenzene	2.2	9.2	151
Dichloroethylene (1,2)	9.7	12.8	57
Diethyl selenide	2.5	—	57
Dimethyl formamide	2.2	—	136
Dioxane	2.0	22.2	54
Ethane	3.1	15.5	—
Ether (diethyl)	1.8	36.5	–49
Ethyl acetate	2.2	11.5	28
Ethyl alcohol	3.3	19.0	54
Ethyl bromide	6.7	11.3	—
Ethyl cellosolve	2.6	15.7	104
Ethyl chloride	4.0	14.8	–58
Ethyl ether	1.9	48	–49
Ethyl formate	2.7	16.5	–4
Ethyl lactate	1.5	—	115
Ethyl nitrite	3.0	50	–31
Ethylene	2.7	28.6	—
Ethylene dichloride	6.2	15.9	56
Ethylene oxide	3.0	80	—
Furfural	2.1	—	140
Gasoline (variable)	1.4–1.5	7.4–7.6	–50
Heptane	1.0	6.0	25
Hexane	1.2	6.9	–15
Hydrogen	4.0	74.2	—
Hydrogen cyanide	5.6	40.0	—
Hydrogen sulfide	4.3	45.5	—
Illuminating gas (coal gas)	5.3	33.0	—
Isobutyl alcohol	1.7	—	82
Isopentane	1.3	—	—
Isopropyl acetate	1.8	7.8	43
Isopropyl alcohol	2.0	—	53
Kerosene	0.7	5	100
Linseed oil	—	—	432
Methane	5.0	15.0	—
Methanol	6.7	36.5	52
Methyl acetate	3.1	15.5	14
Methyl bromide	13.5	14.5	—
Methyl butyl ketone	1.2	8.0	—
Methyl chloride	8.2	18.7	—
Methyl cyclohexane	1.1	—	25
Methyl ether	3.4	18	—
Methyl ethyl ether	2.0	10.1	–35
Methyl ethyl ketone	1.8	9.5	30
Methyl formate	5.0	22.7	–2
Methyl propyl ketone	1.5	8.2	—

TABLE 12-1
Combustibility Characteristics of Selected Pure Gases and Vapors in Air—Continued

Gas or vapor	Lower limit (% by volume)	Upper limit (% by volume)	Closed-cup flash point (°F)
Mineral spirits No. 10	0.8	—	104
Naphthalene	0.9	—	176
Nitrobenzene	1.8	—	190
Nitroethane	4.0	—	87
Nitromethane	7.3	—	95
Nonane	0.83	2.9	88
Octane	0.95	3.2	56
Paraffin oil	—	—	444
Paraldehyde	1.3	—	—
Pentane	1.4	7.8	—
Propane	2.1	10.1	—
Propyl acetate	1.8	8.0	58
Propyl alcohol	2.1	13.5	59
Propylene	2.0	11.1	—
Propylene dichloride	3.4	14.5	60
Propylene oxide	2.0	22.0	—
Pyridine	1.8	12.4	74
Rosin oil	—	—	266
Toluene (toluol)	1.3	7.0	40
Turpentine	0.8	—	95
Vinyl ether	1.7	27.0	—
Vinyl chloride	4.0	21.7	—
Water gas (variable)	6.0	70	—
Xylene (xylol)	1.0	6.0	63

Source: C.R. Bruner.

TABLE 12-2
Auto-Ignition Temperature of Some Common Organic Compounds

Compound	*Temperature (°F)*	*Compound*	*Temperature (°F)*
Acetone	1,000	Hydrogen cyanide	1,000
Ammonia	1,200	Hydrogen sulfide	500
Benzene	1,075	Kerosene	490
Butadiene	840	Maleic anhydride	890
Butyl alcohol	693	Methane	999
Carbon disulfide	257	Methyl alcohol	878
Carbon monoxide	1,205	Methyl ether	662
Chlorobenzene	1,245	Methyl ethyl ketone	960
Cresol	1,038	Mineral spirits (nonspecific)	475
Cyclohexane	514	Nitrobenzene	924
Dibutyl phthalate	760	Oleic acid	685
Dichloromethane	1,185	Petroleum naphtha	475
Ethane	950	Phenol	1,319
Ethyl acetate	907	Phthalic anhydride	1,084
Ethyl alcohol	799	Propane	874
Ethyl benzene	870	Propylene	940
Ethyl chloride	965	Styrene	915
Ethyl ether	366	Sulfur	450
Ethylene dichloride	775	Toluene	1,026
Ethylene glycol	775	Turpentine	488
Ethylene oxide	804	Vinyl acetate	800
Furfural	739	Xylene	924
Furfural alcohol	915		
Gycerin	739		
Hydrogen	1,076		

Source: C.R. Bruner

TABLE 12-3
Physical and Chemical Constants[a] of Common Natural Gas Constituents

Component	Molecular weight	Boiling point (°C)	Specific gravity: Liquid at 16°C/16°C[a]	Specific gravity: Gas[b] (air = 1.00)	Grossa heating value MJ/m^3 [c]
n-Butane	58.124	−0.5	0.5844[e]	2.0068	121.36
Carbon dioxide	44.010	−78.5	0.827[e]	1.5195	—
Ethane	30.070	−88.6	0.3564[e]	1.0382	65.80
Hydrogen sulfide	34.076	−60.3	0.79[e]	1.1765	23.70
Isobutane	58.124	−11.7	0.5631[e]	2.0068	121.01
Isopentane	72.151	27.8	0.6247	2.4911	148.82
Methane	16.043	−161.5	0.3[d]	0.5539	37.56
Nitrogen	28.013	−195.8	0.808[f]	0.9672	—
n-Pentane	72.151	36.1	0.6310	2.4911	148.16
Propane	44.097	−42.0	0.5077[e]	1.5225	93.65
Water	18.015	100.0	1.000	0.6220	—

[a]At 16°C relative to water at 16°C.

[b]At 101.35 kPa (1 atm) and 16°C.

[c]To convert MJ/m^3 to Btu/ft^3, multiply by 26.88.

[d]Apparent value at 16°C.

[e]At saturation pressure and 16°C.

[f]Density of liquid, g/mL, at normal boiling point.

Source: R.E. Kirk et al.

FIGURE 12-7

Approximate Percent Reduction Ranges for Add-On HAP Control Equipment

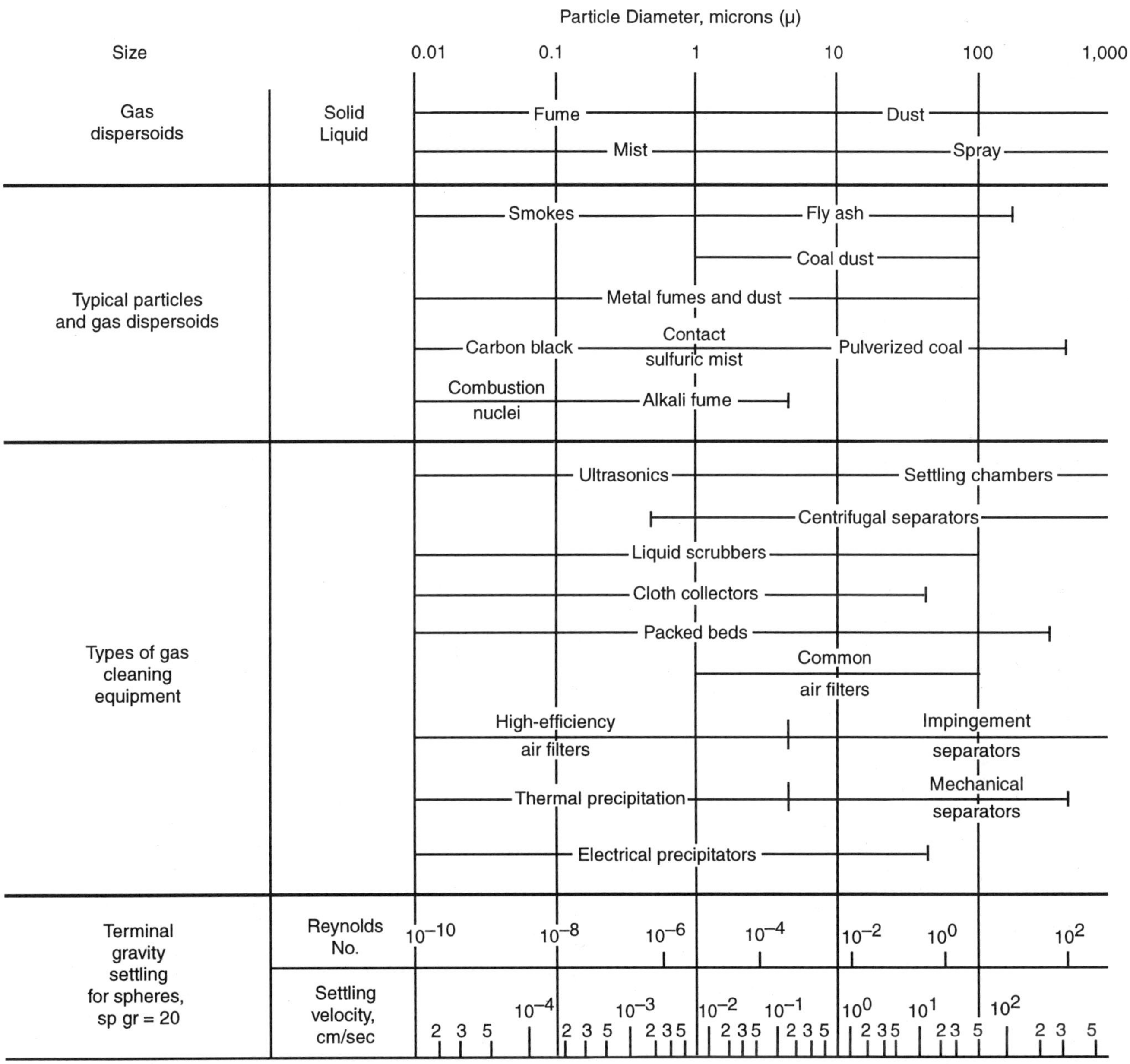

Source: C.R. Bruner.

TABLE 12-4

Key Emission Stream and HAP Characteristics for Selected Control Techniques for Organic Vapors

Control device	Emission stream characteristics					HAP characteristics[a]			
	HAP/ organics content[b] (ppmv)	Heat content (Btu/scf)	Moisture content (%)	Flow rate (scfm) (unless otherwise noted)	Temperature (°F)	Molecular weight (lb/lb mole)	Solubility	Vapor pressure (mm Hg)	Adsorptive properties
Thermal incinerator	>20; (<25% of LEL[c])			<50,000[d]					
Catalytic incinerator	50–10,000; (<25% of LEL[c])			<50,000					
Flare		>300[e]		<2,000,000 lb /hr					
Boiler/process heater[f]		>150g		Steady					
Carbon adsorber	700–10,000; (<25% of LEL[c])		<50%[h]	300–200,000	≤130	45–130			Must be able to adsorb on/ desorb from available adsorbents
Adsorber	250–10,000			1,000–100,000			Must be readily soluble in water or other solvents		
Condenser	>5,000–10,000			<2,000				>10 (at room temperature)	

HAP = hazardous air pollutant; LEL = lower explosive unit; VOC = volatile organic compound.

[a]Refers to the characteristic of the individual HAP if a single HAP is present and to that of the HAP mixture if a mixture of HAPs is present.

[b]Determined from HAP/hydrocarbon content.

[c]For emission streams that are mixtures of air and VOCs; in some cases, the LEL can be increased to 40%–50% with proper monitoring and control.

[d]For packaged units; multiple package or custom-made units can handle larger flows.

[e]Based on EPA's guidelines for 98% destruction efficiency.

[f]Applicable if such a unit is already available on site.

[g]Total heat content.

[h]Relative humidity. Applicable for HAP concentration less than about 1,000 ppmv.

Source: U.S. EPA.

FIGURE 12-8
Approximate Percent Reduction Ranges for Add-on HAP Control Equipment

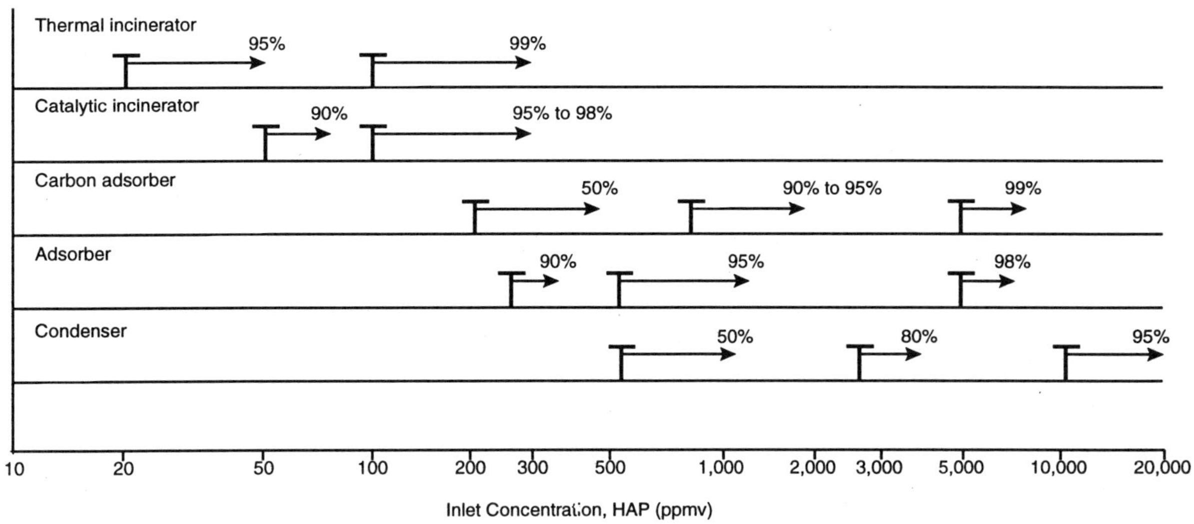

Source: U.S. EPA.

TABLE 12-5
Atmospheric Composition of Air

Components	*% by volume (near ground level)*
Major	
Nitrogen	78.08
Oxygen	20.95
Minor	
Argon	0.934
Carbon dioxide	0.035
Noble gases	
—Neon	1.818×10^{-3}
—Helium	5.24×10^{-4}
—Krypton	1.14×10^{-4}
—Xenon	8.7×10^{-6}
Selected trace gases	
—Methane	1.6×10^{-4}
—Nitrous oxide	3×10^{-5}
—Carbon monoxide	$\sim 1.2 \times 10^{-5}$

Source: Adapted from S.E. Manahan

TABLE 12-6
Emission Testing and Monitoring Methods

Promulgated test methods[1]

Method 1—Traverse Points
Method 1A—Small Ducts
Method 2—Velocity (S-Type Pitot)
Method 2A—Volume Meters
Method 2B—Exhaust Volume Flow Rate
Method 2C—Standard Pitot
Method 2D—Rate Meters
Method 2E—Landfill Gas Production Flow Rate
Method 3—Molecular Weight
Method 3A—CO_2 (instrumental analyzer procedure)
Method 3B—CO_2 (orsat analyzer procedure)
Method 3C—CO_2 (thermal conductivity detector)
Method 4—Moisture Content in Stack Gases
Method 5—PM
Method 5A—PM-Asphalt Roofing (industry)
Method 5B—PM-Nonsulfuric Acid
Method 5D—PM-Baghouses (particulate matter)
Method 5E—PM-Fiberglass Insulation Industry
Method 5F—PM-Nonsulfate
Method 5G—PM-Wood Heaters From a Dilution Tunnel
Method 5H—PM-Wood Heaters From a Stack
Method 6—Sulfur Dioxide (SO_2)
Method 6A—SO_2
Method 6B—SO_2 - Long-term integrated procedure
Method 6C—SO_2 - Instrumental analyzer procedure
Method 7—NO_x
Method 7A—NO_x - Ion Chromatographic Method
Method 7B—NO_x - Ultraviolet Spectrophotometry Method
Method 7C—NO_x - Colorimetric Method
Method 7D—NO_x - Ion Chromatographic Method
Method 7E—NO_x - Instrumental Analyzer Procedure
Method 8—Sulfuric Acid Mist
Method 9—Visual Opacity Determination
Method 10—Carbon Monoxide (CO) Nondispersive Infrared Analyzer (NDIR) Method
Method 10A—CO for Certifying CEMS
Method 10B—CO From Stationary Sources
Method 11—H_2S - Content of Fuel in Petroleum Refineries
Method 12—Inorganic Lead
Method 13A—Total Fluoride
Method 13B—Total Fluoride (Specific Ion Electrode)
Method 14—Fluoride for Primary Aluminum Plants
Method 14A—Total Fluoride Emissions from Selected Sources at Primary Aluminum Plants
Method 15—H_2S
Method 15A—Total Reduced Sulfur
Method 16—Sulfur (Semicontinuous Determination)
Method 16A—Total Reduced Sulfur (Impinger method)
Method 16B—Total Reduced Sulfur (gas chromatography analysis)
Method 17—In-Stack Particulate
Method 18—VOC by GC
Method 19—SO_2 Removal and Efficiency
Method 20—NO_x From Stationary Gas Turbines
Method 21—VOC Leaks
Method 22—Fugitive Opacity
Method 23—Dioxin and Furan
Method 24—Surface Coatings
Method 24A—Printing Inks and Related Coatings
Method 25—Gaseous Nonmethane Organic Emissions
Method 25A—Gaseous Organic Concentration (flame ionization)
Method 25B—Gaseous Organic Concentration (infrared analyzer)
Method 25C—Nonmethane Organic Carbon in Landfill Gases
Method 25D—VOC of Waste Samples
Method 25E— Vapor Phase Organic Concentration in Waste Samples
Method 26—Hydrogen Chloride
Method 26A—Hydrogen Halide & Halogen (isokinetic method)
Method 27—Vapor Tightness of Gasoline Tank (pressure vacuum)
Method 28—Certification and Auditing - Wood Heaters
Method 28A—Air-to-Fuel Ratio
Method 29—Metals Emissions From Stationary Sources
Method 101—Mercury From Chlor-Alkali Plants (air streams)
Method 101A—Mercury From Sewage Sludge Incinerators
Method 102—Mercury From Chlor-Alkali Plants (hydrogen streams)
Method 103—Beryllium Screening Method
Method 104—Beryllium Emissions Determination
Method 105—Mercury in Wastewater Treatment Plant Sewage Sludge

TABLE 12-6
Emission Testing and Monitoring Methods—Continued

Promulgated test methods—Continued

- Method 106—Determination of Vinyl Chloride
- Method 107—Vinyl Chloride Content of Inprocess Wastewater Samples
- Method 107A—Vinyl Chloride Content of Solvents
- Method 108—Particulate and Gaseous Arsenic Emissions
- Method 108A—Determination of Arsenic Content in Ore Samples From Nonferrous Smelters
- Method 108B—Arsenic (>10 μg)
- Method 108C—Arsenic (<10 μg)
- Method 111—Polonium-210 Emissions
- Method 114—Radionuclide Emissions
- Method 115—Radon-222 Emissions
- Method 201—PM-10 (In-Stack) - Exhaust gas recycle procedure
- Method 201A—PM-10 (In-Stack) - Constant sampling rate procedure
- Method 202—Condensable Particulate Matter
- Method 204—Permanent or Temporary Total Enclosure (TTE) for Determining Capture Efficiency
- Method 204A—VOCs in Liquid Input Stream
- Method 204B—VOCs in Captured Stream
- Method 204C—VOCs in Captured Stream (dilution technique)
- Method 204D—Fugitive VOCs From Temporary Total Enclosure
- Method 204E—Fugitive VOCs From Building Enclosure
- Method 204F—VOCs in Liquid Input Stream (distillation)
- Method 205—Gas Dilution Calibration
- Method 301—Visible Emissions From Field Validation of Pollutant Measurement Methods
- Method 303—Visible Emissions From By-Product Coke Oven Batteries
- Method 303A—Nonrecovery Coke Oven Batteries
- Method 304A—Biodegradation Rates - Vent Option
- Method 304B—Biodegradation Rates - Scrubber Option
- Method 305—Potential VOCs in Waste
- Method 306—Chromium Emissions - Electroplating/Anodizing Operations
- Method 306A—Chromium Emissions - Electroplating/Anodizing (mason jar method)
- Method 306B—Surface Tension for Tanks - Electroplating/Anodizing Operations
- Method 307—Emissions From Solvent Vapor Cleaners
- Method 308—Methanol Emissions
- Method 310A—Residual Hexane
- Method 310B—Residual Solvent
- Method 310C—Residual N-Hexane in EDPM Rubber
- Method 311- HAPs in Paints and Coatings
- Method 312A—Styrene in SBR Latex (GC)
- Method 312B—Styrene in SBR Latex by Capillary GC
- Method 312C—Styrene in SBR Latex Produced by Emulsion Polymerization
- Method 313A—Residual Hydrocarbon in Rubber Crumb
- Method 313B—Residual Hydrocarbon in Rubber Crumb by Capillary GC
- Method 315—PM and MCEM from Aluminum Production Facilities

Proposed test methods[2]

- Proposed Method 23—5/31/95 FR Notice (Proposed Revision)
- Proposed Method 203—Continuous Opacity Monitoring for Compliance
- Proposed Method 203A—Visual Opacity (time-averaged)
- Proposed Method 203B—Visual Opacity (time-exception)
- Proposed Method 203C—Visual Opacity (instantaneous limitation)
- Proposed Method 308—Procedure for Methanol Emissions
- Proposed Method 316—Sample and Analysis for Formaldehyde emissions in the Mineral Wool and Wool Fiberglass Industries
- Proposed Method 318—Extractive FTIR Method for Measurement of Emissions From the Mineral Wool and Wood Fiberglass Industries
- Proposed Method 319—Filtration Efficiency for Paint Overspray Arrestors
- Proposed Method 320—Vapor Phase Organic and Inorganic Emissions by Extractive FTIR
- Proposed Method 321—Gaseous HCl Emissions at Portland Cement Kilns by FTIR
- Proposed Method 322—HCl Emissions From Portland Cement Kilns by GFCIR

Approved alternative test methods[3]

- ALT-001—SO_2 Interference in Methods 7 and 7A
- ALT-002—Evaluation of Gas Tanks for CO Analysis
- ALT-003—Potential Safety Hazard in Method 101A
- ALT-004—Alternative for Methods 3A and 6C
- ALT-005—Acceptability of Teflon Bags (Revised)

TABLE 12-6
Emission Testing and Monitoring Methods—Continued

Approved alternative test methods[3]—Continued

- ALT-006—ICP Alternative for Methods 12
- ATL-007—Use of Dilution Probes - Method 6C
- ALT-008—Alternative Method-Moisture Midget Impingers
- ALT-009—Alternative Method 5 Post-Test Calibration
- ALT-010—Alternatives for Method 11
- ALT-011—Alternative Method 2 Thermocouple Calibration
- ALT-012—Alternative Stack Gas Volume Flow Rate
- ALT-013—Alternative for Method 20
- ALT-014—Methods 306 and 306A, Section 5.2.3 Decorative and Hard Chromium Electroplating and Anodizing Operations
- ALT-015—3-D Pitot Method for Velocity and Flow Rate
- ALT-016—Optical Scintillation Anemometers for Aluminum Smelter Potroom Roof Monitor Effluent Velocity
- ALT-017—Determination of Gaseous Organic Compounds by Direct Interface GC/MS

Conditional test methods[4]

- CTM 1—Butadiene Method
- CTM 2—Particulate (PM Screening) Method
- CTM 3—Particulate (PM Modified High Volume) Method
- CTM 8—Conditional Test Method for Acrylonitrile
- CTM 10—Perchloroethylene in Wet Waste (Filters and Still Bottoms)
- CTM 11—Halogenated Organics Method
- CTM 14—Benzene Matter Emissions
- CTM 19—Gas Velocity and Volumetric Flowrate Under Cyclonic Flow Conditions
- CTM 22—Nitric Oxide
- CTM 23—Draft Sampling Method for Isocyanates
- CTM 24—Draft Analysis for Isocyanates by High Performance Liquid Chromatography
- CTM 25—Microbial Survivability Test for Medical Waste Incinerator Ash
- CTM 26—Microbial Survivability Test for Medical Waste Incinerator Emissions
- CTM 27—Procedure for Collection and Analysis of Ammonia in Stationary Sources
- CTM 28—Direct Interface GC/MS Performance-Based Method
- CTM 29—CARB Method 435-Asbestos in Serpentine Aggregate
- CTM 30—GRI Method for Determination of NO_x

Preliminary test methods[5]

- PRE 1—Draft Method for Direct Measurements of Gas Velocity and Volumetric Flowrate Under Cyclonic Flow Conditions
- PRE 2—Particulate (PM Screening) Method
- PRE 3—DRAFT Ontario Hydro Method for Mercury in Flue Gas Generated From Coal-Fired Stationary Sources
- PRE 4—DRAFT Method for Determination of PM-10/PM-2.5

CARB = California Air Resources Board; CEMS = continuous emissions monitoring system; CO = carbon monoxide; CO_2 = carbon dioxide; *FR* = *Federal Register*; FTIR = Fourier transfer infrared spectroscopy; GC = gas chromatography; GFCIR = infrared gas-filter correlation; GRI = Gas Research Institute; HCl = hydrogen chloride; H_2S = hydrogen sulfide; ICP = inductively coupled plasma; MCEM = methylene chloride extractable materials; MS = mass spectroscopy; NO_x = nitrogen oxides; PM = particulate matter; PM-2.5 = particulate matter with an aerodynamic diameter less than or equal to 2.5 microns; PM-10 = particulate matter with an aerodynamic diameter less than or equal to 10 microns; SO_2 = sulfur dioxide; VOC = volatile organic compound.

[1]These test methods are promulgated and directly cited by specific regulations for determining compliance under 40 *CFR* Part 51, Appendix M; Part 60, Appendix A; Part 61, Appendix B; and Part 63, Appendix A. (They are identified by their abbreviated titles and may be used without further EPA approval.)

[2]These test methods have been published in the *Federal Register* as proposed rules but have not yet been finalized. Proposed methods may be new methods or revisions to existing promulgated methods and are intended for review purposes only prior to promulgation.

[3]These test methods are approved alternatives to the methods in 40 *CFR* Parts 51, 60, 61, and 63.

[4]These test methods have been evaluated by EPA and may be applicable to one or more categories of stationary sources. Methods in this category *must be approved as alternatives* prior to use.

[5]The performance of these methods is not as well defined as the conditional methods. However, they may be useful in limited applications until more supporting information is available. As in the case with conditional test methods, all procedures in this category *must be approved as alternative methods* prior to use.

Source: U.S. EPA.

TABLE 12-7
Common Conversion Factors

Length

1 angstrom = 10^{-8} cm = 10^{-10} meters (m)

1 micron = 10^{-3} millimeter (mm) = 10^{-6} m

1 inch = 2.54 centimeters (cm)

1 ft = 0.305 m

1 m = 3.281 ft

1 mile = 5,280 ft = 1,609 m

Mass

1 lb = 453.6 grams (g) = 0.454 kilograms (kg)

1 lb = 7,000 grains = 16 oz

1 ton = 2,000 lb

1 tonne (metric ton) = 2,200 lb

1.1 ton = 1 megagram (Mg)

Time

1 hr = 3,600 sec

1 work year ≅ 2,000 hours per person

1 year = 8,760 total hours

Energy

1 erg = 1 dyne cm

1 joule (J) = 10^7 erg

1 calorie (cal) = 4,185 J

1 cal/g mole = 1.8 Btu/lb mole

1 Btu = 252.2 cal = 777.5 ft lb_f

1 kilowatt-hour (KWh) = 3,412 Btu
= 1.341 horsepower (hp)/hr = 3.6×10^6 J

1 hp hr = 1.98×10^6 ft lb_f

1 ton refrigeration = 1,200 Btu/hr

Power

1 watt = 1 J/sec = 10^7 erg/sec

1 hp mechanical ≅ 746 watt = 33,000 ft lb_f/min
≅ 2,545 Btu/hr

1 hp (boiler) ≅ 34.51 lb steam/hr

Density

air = 1.20×10^{-3} g/cm^3
= 7.49×10^{-2} lb/ft^3 at SC[b]

H_2O = 1 g/cm^3 = 8.34 lb/gal at 4°C = 62.4 lb/ft^3

Hg = 13.6 g/cm^3 = 849 lb/ft^3 at 4°C

glass ≅ 2.2 g/cm^3 ≅ 137.3 lb/ft^3

earth ≅ 1.5 g/cm^3 ≅ 93.5 lb/ft^3

coal ≅ 0.72 – 1 g/cm^3 = 45 – 62.4 lb/ft^3

fly ash ≅ 1.0 g/cm^3 = 62.4 lb/ft^3

1 g/standard cubic foot (scf)
= 2,290 mg/normal cubic meter (Nm^3)[a]

1 g/Nm^3 = 0.438 g/scf

1 grain/ft^3 = 2.29 g/m^3

Viscosity

1 slug/(ft sec) = 32.174 lbm/(ft/sec)

1 poise = 1 g/(cm sec) = 100 centipoise (cp)
= absolute viscosity

1 cp = 0.000672 lb/(sec ft)

1 stoke = cm^2/sec = poise/unit density = kinematic viscosity

air = 1.83×10^{-4} g/(cm sec) at SC[b]

water = 1 cp = 0.01 g/(cm sec) at SC[b]

1 cp = 6.72×10^{-4} lb_m/(sec ft) = 2.42 lb_m/(hr ft)
= 0.01 g/(cm sec) = 2.089×10^{-5} lb_f sec/ft^2

Temperature

Kelvin (K) = degrees Celsius (°C) + 273.16

Degrees Fahrenheit (°F) = (1.8°C × °C) + 32

Degrees Rankine (°R) = °F + 459.49

°C = (°F – 32) × 0.55

Speed

1 ft/sec = 30.48 cm/sec = 0.6818 mi/hr

1 mi/hr = 0.447 m/sec

Velocity

1 scfm = 1.7 Nm^3/hr = 28,317 cm^3/min

1 gpm = 0.137 ft^3/min = 0.00378 m^3/min

Acceleration

gravity (g) = 32.174 ft/sec^2
= 980.665 cm/sec^2 at sea level

Force

1 dyne = 1 g cm/sec^2

1 newton (N) = 1 kg m/sec^2 = 10^5 dynes

TABLE 12-7
Common Conversion Factors—Continued

Pressure

1 atm = 1.01325 bar = 14.696 lb/in^2
= 29.92 in Hg (32°F)
= 760 mm Hg (32°F)
= 33.94 ft H_2O (60°F)
= 760 torr

1 lb/in^2 = 0.014 in Hg (32°F)

1 pascal (Pa) = 1 newton/m^2

1 bar = 10^5 Pa

Geometry

circle area = $\pi r^2 = \pi d^2/4$

circle circumference = πd

cylinder volume = $\pi r^2 \times$ height

sphere area = $4\pi r^2 = \pi d^2$

sphere volume = $4/3\pi r^3 = 1/6\pi d^3$

Area

1 m^2 = 1.2 yd^2

1 acre = 43,560 ft^2 = 0.00156 mi^2

1 hectare = 2.5 acres

1 km^2 = 0.386 mi^2

Volume

1 ft^3 = 7.48 U.S. gallon (gal)
= 28.32 liter (L) = 0.02832 m^3

1 gal = 3.785 L

1 barrel = 42 gal (U.S.) oil

1 L = 1.057 quarts

1 L = 0.2642 gal

1 cm^3 = 1 mL at 4°C

1 m^3 = 35.31 ft^3 = 1,000 L

Miscellaneous constants

g_c = 32.174 lb$_m$ ft/(lb$_f$ sec^2) = 980.7 cm/sec^2
= 1 kg m/(N sec^2)

Natural log base e = 2.7183

Natural log (ln) 10 = 2.3026

$\pi \cong 3.14159$

Boltzmann's constant = k = 1.38 × 10^{-16} g cm^2/(sec^2 molecule Kelvin [K]) = 1.38 × 10^{-23} joule/K

Avogadro's number = 6.02 × 10^{23} atoms/mole

gas constants (R) = 82.05 atm cm^3/(g mole K)
= 1,544 psf ft^3/lb mole R

1 lb mole = 359 ft^3 (ideal gas at STP[c])

1 g mole = 22.4 L (ideal gas at STP[c])

Convert parts per million (ppm) volume to mass per volume at normal conditions:

$$\text{ppm} = \frac{\text{mg/m}^3 \times 24.5}{\text{molecular weight}}$$

$$\text{ppm} = \frac{\mu\text{g/m}^3 \times 24.5}{\text{molecular weight} \times 1{,}000}$$

Note: molecular weight is that of air contaminant of interest.

Btu = British thermal unit; H_2O = water; Hg = mercury; SC = standard conditions.

[a]At normal conditions of 25°C and 1 atm.

[b]Standard conditions of 20°C and 1 atm.

[c]Standard temperature and pressure (STP) of 0°C and 1 atm.

Source: Elsevier Science Inc.

13 Keyword Index

Keyword	Explanatory text

@Elsevier Science Inc. 0-444-10032-6/98/$0+$19.00+$5.50

14 Glossary of Terms

A&WMA	Air & Waste Management Association
ACT	Alternative control technique
ADI	Applicability Determination Index
AFR	Automated Form R
AIChE	American Institute of Chemical Engineers
AIRS	Aerometric Information Retrieval Systems
ALAPCO	Association of Local Air Pollution Control Officials
AMTIC	Ambient Monitoring Technology Information Center
ANPRM	Advanced notice of proposed rulemaking
APC	*The Air Pollution Consultant*
BAC	Best available control
BACT	Best available control technology
BDT	Best demonstrated technology
CAA	Clean Air Act
CARB	California Air Resources Board
CAS	Chemical Abstract Service
CATC	Clean Air Technology Center
CFC	Chlorofluorocarbon
CFR	*Code of Federal Regulations*
CHIEF	Clearinghouse for Inventories and Emission Factors
CLAM	Clean Air Marketplace
CMS	Continuous monitoring system
CO	Carbon monoxide
COMS	Continuous opacity monitoring system
CTC	Control Technology Center
CTG	Control techniques guideline
EAB	Environmental Appeals Board
EFIG	EPA's Emission Factor and Inventory Group
EMA	Emissions Marketing Association
EMS	Environmental Management System
EMTIC	Emission Measurement Technical Information Center
EPCRA	Emergency Planning and Community Right-to-Know Act
FIP	Federal implementation plan
FIRE	Factor information and retrieval system
FR	*Federal Register*
FTIR	Fourier transform infrared spectroscopy
GC	Gas chromatograph
GILS	Government Information Locator Service
GPO	Government Printing Office
HAP	Hazardous air pollutant
HBFC	Hydrobromofluorocarbon
HCFC	Hydrochlorofluorocarbon
HON	Hazardous organic NESHAP
IPPF	Independent power production facility
IRIS	Integrated Risk Information System
ISO	International Organization for Standardization
L&E	Locating and estimating
LAER	Lowest achievable emission rate
LDAR	Leak detection and repair
MACT	Maximum achievable control technology
MSW	Municipal solid waste
MVAC	Motor vehicle air conditioner
MWC	Municipal waste combustor
NAAQS	National ambient air quality standards
NCEA	National Center for Environmental Assessment
NCEPI	National Center for Environmental Publications and Information
NESHAP	National emission standards for hazardous air pollutants
NET	Net emission trends
NMOC	Nonmethane organic compound
NO_2	Nitrogen dioxide
NO_x	Nitrogen oxides
NPRM	Notice of proposed rulemaking
NSO	Nonferrous smelter orders
NSPS	New source performance standards
NSR	New source review
NTIS	National Technical Information Service
NTP	National Toxicology Program
O_3	Ozone
OAP	Office of Atmospheric Programs
OAQPS	Office of Air Quality Planning and Standards
OAR	Office of Air & Radiation
ODS	Ozone-depleting substances
OECA	Office of Enforcement and Compliance Assurance
OLS	Online Library System
OMB	Office of Management and Budget
OPPT	Office of Prevention, Pesticides, and Toxic Substances
OSHA	Occupational Safety and Health Administration
.pdf	Portable document format

PM-2.5	Particulate matter with an aerodynamic diameter less than or equal to 2.5 microns
PM-10	Particulate matter with an aerodynamic diameter less than or equal to 10 microns
PSD	Prevention of significant deterioration
RAB	American Society of Quality Registrar Accreditation Board
RACT	Reasonably available control technology
RBLC	RACT/BACT/LAER Clearinghouse
RCRA	Resource, Conservation, and Recovery Act
RMP	Risk management plan
SCC	Source classification code
SIC	Standard Industrial Classification
SIP	State implementation plan
SNAP	Significant new alternatives policy
SO_2	Sulfur dioxide
SO_x	Sulfur oxides
SOCMI	Synthetic organic chemical manufacturing industry
STAPPA	State and Territorial Air Pollution Program Administrators
TQ	Threshold quantity
TTN	Technology Transfer Network
TTNWeb	Technology Transfer Network Web site
UATW	Unified Air Toxics Website
URL	Uniform resource locator
USC	United States Code
VOC	Volatile organic compound